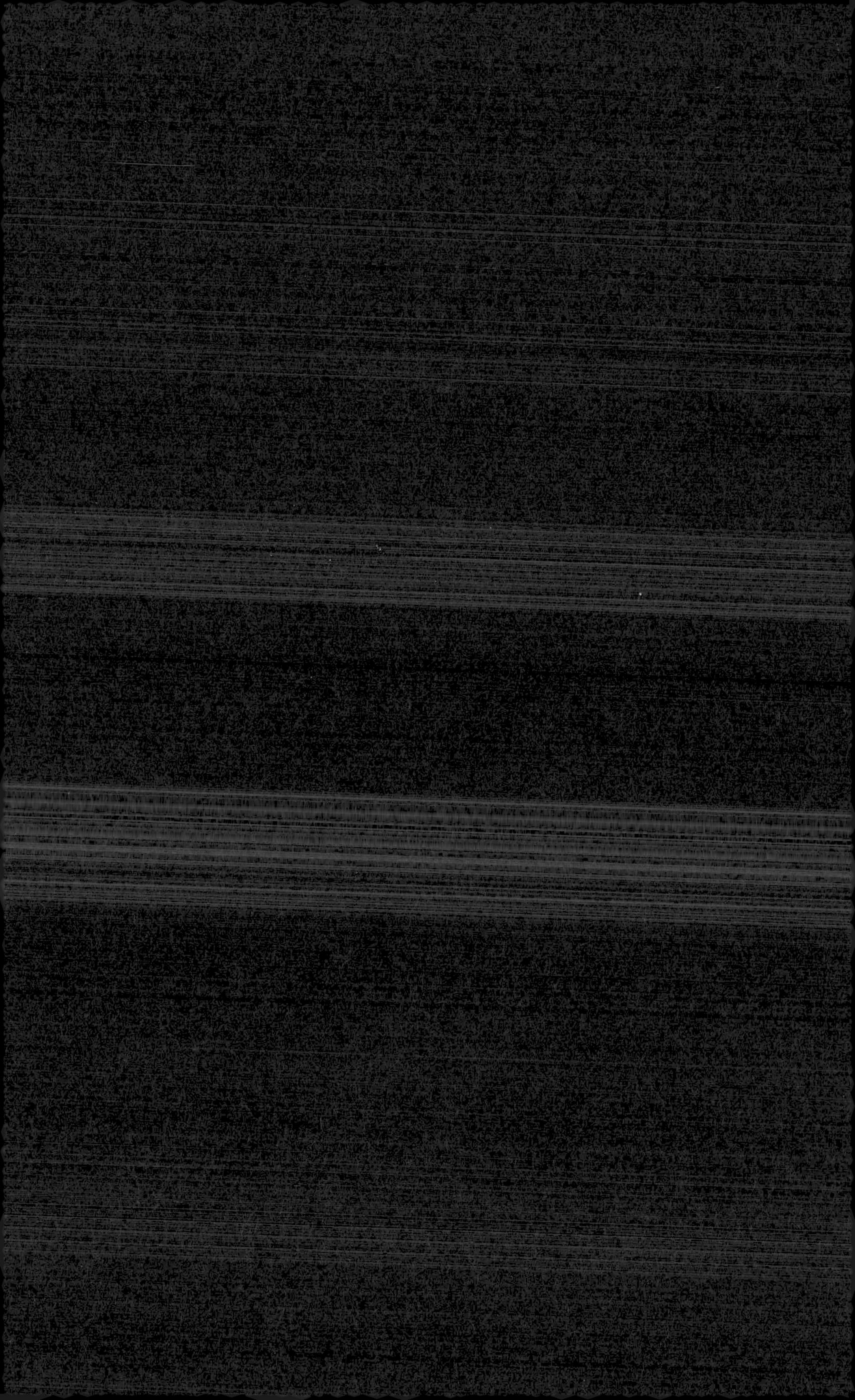

Technocolonialism

For John and Alex

Technocolonialism

When Technology for Good is Harmful

Mirca Madianou

polity

First published in 2025 by Polity Press

Polity Press
65 Bridge Street
Cambridge CB2 1UR, UK

Polity Press
111 River Street
Hoboken, NJ 07030, USA

ISBN-13: 978-1-5095-5902-2 (hardback)
ISBN-13: 978-1-5095-5903-9 (paperback)

A catalogue record for this book is available from the British Library.

Library of Congress Control Number: 2024935065

Typeset in 11.5 on 14 Adobe Garamond
by Fakenham Prepress Solutions, Fakenham, Norfolk NR21 8NL
Printed and bound in Great Britain by CPI Group (UK) Ltd, Croydon

For further information on Polity, visit our website:
politybooks.com

Contents

Abbreviations

AAP	Accountability to Affected People
AI	Artificial Intelligence
AI4SG	Artificial Intelligence for Social Good
ANN	Artificial Neural Network
BIMS	Biometric Identity Management System
CFM	Complaint and Feedback Mechanisms
CHS	Core Humanitarian Standard on Quality and Accountability
CIA	Central Intelligence Agency (USA)
CSR	Corporate Social Responsibility
CWC	Communicating with Communities
DRC	Danish Refugee Council
EU	European Union
EURODAC	European Dactyloscopy System
FA	Forensic Architecture
GDPR	General Data Protection Regulation (EU)
GIS	Geographic Information System
HAP	Humanitarian Accountability Partnership
ICE	Immigration and Customs Enforcement (USA)
ICRC	International Committee of the Red Cross
IFRC	International Federation of Red Cross and Red Crescent Societies
INGO	International Non-governmental Organization
IRC	International Rescue Committee
MSF	Médecins Sans Frontiers
NLG	natural language generation
NLU	natural language understanding
OIOS	UN Office of Internal Oversight Services
OPPAR	Office of the Presidential Assistant for Rehabilitation and Recovery (Philippines)

PRIMES	Population Registration and Identity Management Ecosystem
SDGs	Sustainable Development Goals
SRAD	Syrian Refugee Affairs Directorate
TPM	Third-Party Monitoring
UN	United Nations
UNHCR	United Nations High Commissioner for Refugees
UNOCHA	United Nations Office for the Coordination of Humanitarian Affairs
WASH	Water, Sanitation and Hygiene Cluster
WFP	World Food Programme

Introduction

Digital Humanitarianism as Technocolonialism

Between 2017 and 2018 over one million Rohingya refugees arrived in Bangladesh after fleeing genocide in Myanmar. In 2018 the office of the United Nations High Commissioner for Refugees (UNHCR) together with the Bangladesh government and a private contractor organized a digital identity programme aiming to collect the biometric data of all refugees. Biometric registrations have become standard in the humanitarian response to refugee flows: one of the first things that happens when a refugee comes into contact with UNHCR is to give their biometric data. The Rohingya registration was controversial and led to protests in the camps. The activists protested for two reasons: first, the digital identity system did not use their own ethnicity, but referred to them as 'Myanmar nationals', which was perceived as a form of symbolic erasure. Second, the Rohingya were concerned that their digital records would be shared with the Myanmar government, thus endangering them in the future. In November 2018 the Rohingya protests culminated in a strike – a sign of resistance. The Bangladeshi government responded forcefully by sending in the police who beat up the protestors. The next day the biometric registrations continued as normal.

The above example encapsulates both the pervasiveness of, and the violence associated with, digital technologies in the humanitarian sector. With over 108 million refugees globally in 2022,[1] and over 339 million people in need of humanitarian assistance in 2023, the sector is facing significant challenges.[2] As humanitarian emergencies and climate disasters become more common, digital innovation and artificial intelligence (AI) are championed as solutions to the complex problems of the sector.

We are witnessing a rapid digitalization and datafication of humanitarian operations. Biometric technologies are not just used in refugee registrations, but have multiple applications across the sector. Blockchain technology underpins virtual cash transfers while cryptocurrencies are being piloted for cash disbursement.[3] Feedback channels are increasingly digitized while AI-powered chatbots are routinely used for communicating with affected communities. Algorithms are employed to decide about who will be included in aid distribution,[4] while AI is being explored to predict refugee flows and future crises.[5] The Covid-19 pandemic accelerated the implementation of digital interventions, as the need to deliver aid remotely increased due to public health measures involving social distancing. Some of these technologies are still at an experimental stage of development with no clear understanding of their potential consequences. The term 'digital humanitarianism' is often used as a shorthand for the uses of digital innovation and data in humanitarian emergencies. Several of these innovations are presented as part of the wider phenomenon of 'technology for social good', or 'AI for social good'. 'Technology for good' initiatives proclaim the benevolent uses of digital technology and AI to address social, economic and environmental challenges.

I first became aware of the phenomenon 'digital humanitarianism' in the hours following the landfall of Super Typhoon Haiyan (locally named Yolanda), which hit the Philippine archipelago on 8 November 2013 and remains one of the strongest storms to ever make landfall. I had recently completed an ethnography of migrant Philippine families and their use of communication technologies to enact family life at a distance. With many friends and colleagues in the country, I was following the news about the storm with concern. Within hours of Typhoon Haiyan making landfall, and while all communications were severed across the affected areas, the web was flooded with optimism about the potential of digital technologies to mitigate the impact of the disaster. Digital volunteers – self-styled 'digital humanitarians' – organized mapathons, which are coordinated events during which members of the public update online maps with information about storm damage using satellite imagery. Hackathons were set up to identify technological solutions while hashtags such as '#digitalJedis' were used on social media.

The destruction caused by the Typhoon was immense. Over 6,300 people died while millions of people were displaced or otherwise affected.[6] The humanitarian response to Haiyan was massive. Hundreds of non-governmental organizations (NGOs) arrived in the affected areas in order to help. UN agencies had a large presence in Tacloban. In total,

3.7 million people received food assistance and over 1.4 million cash assistance; about 1.3 million people were given access to safe water and about 570,000 households were housed in emergency shelters.[7] These are just some indicative figures that reveal the complexity of the relief efforts.

The response to Haiyan was quickly branded the most accountable humanitarian response to date. The high levels of mobile phone connectivity in the Philippines were seen as a 'laboratory opportunity' for the implementation of digitized accountability projects, as one of my interlocutors from the aid sector later put it. I became curious to find out whether this newfound enthusiasm for digital technologies lived up to the expectations. More importantly, how did the communities affected by the Typhoon respond to digital innovation and how did they use technologies themselves?

Together with several colleagues, Jonathan Ong, Liezel Longboan, Jayeel Cornelio and Nicole Curato, I spent a year following the politics of recovery in the aftermath of Typhoon Haiyan. Through our ethnography we shared the experiences of affected communities and their communication practices. Apart from participant observation and digital ethnography we also interviewed local residents displaced or otherwise affected by the disaster, as well as humanitarian officers, government officials, volunteers and activists. Through this research I started following the trails of data and innovation in the humanitarian system. Who produces the data, how is it used and where does it end up? What do the data trails reveal about the power geometries of humanitarianism?

The Haiyan project led me to a wider exploration of the uses of data and technology in the humanitarian sector more broadly. The start of the refugee crisis in 2015 was accompanied by a similar avalanche of hackathons in support of refugees, as had happened in the aftermath of Typhoon Haiyan. But one event in particular revealed to me how digital platforms were implicated in the containment of the humanitarian crisis and ended up amplifying the harms of the asylum system. In summer 2016, while visiting Greece, I became aware that the pre-registration process for asylum applications took place by Skype. Pre-registration with the Greek authorities is a pre-requisite for the asylum application process to even begin. The rationale was that online interviews would minimize travel and waiting time and streamline the application process.

Conversations with NGO officers and refugee representatives revealed a different side to the story. The slots for these online interviews were scarce, especially for people of certain nationalities or languages who had to compete to get an appointment within the one hour per week allocated to their language group. For example, there were significantly fewer slots for refugees speaking Pashto than Arabic. The online system meant there was no person to speak with directly to explain one's circumstances and why it was impossible to book a pre-registration appointment. Delays of several months, or in some cases, years were reported while refugees were left in limbo.[8]

Between 2016 and 2021 I continued to research the uses of digital technologies and computation in the humanitarian sector. During this time the technologization of humanitarian operations intensified, while imaginaries about AI and its potential to solve some of the challenges of the sector became prominent. Through interviews and fieldwork with humanitarian workers, donors, digital developers, entrepreneurs, private companies, hackers as well as affected communities themselves I explored the design, uses and consequences of data and AI in emergencies. I also conducted digital ethnography and participant observation in spaces of innovation such as industry events, workshops, hackathons and mapathons. By following 'the thing' (Marcus, 1995), I was able to trace the flows of data and innovation – and the discourses around data, innovation and AI – across the sector.

The book draws on these two projects over the course of ten years. I elaborate on the two projects and the data collected in the 'Note on Methods' at the end of the book. I combine research material from these two projects throughout the book as well as additional desk research between 2021 and 2023. When drawing on the ethnography of the aftermath of Typhoon Haiyan I refer to the 'Haiyan project'. When drawing on the subsequent research on the uses of digital technologies and AI in the humanitarian space, I refer to the 'digital humanitarianism project'. The purpose of the book is not to report in detail from these two projects, which has been done in article publications, but to synthesize the material into new theory. In combination, the two projects illuminate different facets of humanitarianism, from the response to disaster emergencies to protracted crises of displacement, and how digital technologies and infrastructures have become central to relief operations.

The projects have enabled me to connect different levels of analysis from top level policy and the design of technologies to their implementation and actual consequences for people living in precarity. Tracing the trails of data revealed the power relations within the humanitarian sector. I argue that the multi-sited optics adopted here is necessary for understanding global crises, digital and institutional infrastructures and their overlaps in the global neoliberal context.

In order to make sense of the practices of data, digital innovation and AI in the humanitarian space I put forward the term 'techno-colonialism'. I use this term to refer to the way that digital innovation, data and AI practices entrench power asymmetries and engender new forms of structural violence and new inequities between the global South and North. Technocolonialism illuminates the convergence of digital developments with humanitarian structures, state power and market forces and the extent to which they reinvigorate and rework colonial relationships.

To support my analysis, I draw on postcolonial and decolonial theory as well as work in social anthropology, critical race studies and the Black radical tradition that argues for the endurance of colonial structures. As technological systems increasingly underpin key humanitarian practices, I turn to infrastructure studies – an interdisciplinary area informed by science and technology studies, information science, social anthropology and geography – to make sense of technology, data and AI as a process of 'infrastructuring'. The fields of critical algorithm / AI studies also inform my understanding of the constituent technologies that become part of the digital infrastructure ecosystem. In Chapter 1, I analyse humanitarianism and its relationship with empire, capitalism, technology and the state. I do so by teasing out the six logics that drive digital developments in the sector. But first, let me explain why colonialism is a relevant framework for understanding digital and AI practices in contemporary humanitarian operations.

Colonialism

Colonialism did not end with the political independence of postcolonial states.[9] It has deep roots and manifests itself in all aspects of public and private lives globally. The fields of postcolonial studies and decolonial theory have made significant contributions to our understanding of the ongoing harms of colonialism. Postcolonial and decolonial theory are often treated as separate or competing approaches. The emergence of postcolonial theory has been associated with the work of Edward Said, Gayatri Spivak and Homi Bhabha among many others. Several, although certainly not all, postcolonial studies scholars originate from the Middle East and South Asia and refer to these regions and their nineteenth-century colonization in their writings. Decolonial theory is associated with the work of Aníbal Quijano, María Lugones, Silvia Rivera Cusicanqui and Walter Mignolo among many others, who focus on South America over a longer period of European colonization from the fifteenth century (Bhambra, 2017). Despite differences in intellectual agendas, traditions (postcolonial theory emerged through literary criticism while decolonial theory through sociology) and geographic origin, I argue that there are productive overlaps and convergences. I draw on both traditions here as well as the Black radical tradition, critical border studies and anthropology.

Quijano's notion of the 'coloniality of power' offers an important approach for understanding the persistence of colonial structures. According to Quijano (2000) the dominance of Eurocentric knowledge, the codification of racial and social discrimination and the pervasiveness of global capitalism explain the ongoing subjugation of the colonized after direct colonial rule. Racial and social orders outlive direct colonization and organize the distribution of material and epistemic resources, which in turn reproduce and naturalize racial orders. The notion of the 'coloniality of power' contributes to our understanding of colonialism not only as a form of expropriation, but as a hegemonic system of Eurocentric knowledge. Eurocentrism here is not just a geographical category: 'it is about the control of knowledge and subjectivities' (Mignolo and Walsh, 2018: 125). Coloniality emerges as the 'complex system of management and control which underlies Western civilization' (Mignolo and Walsh, 2018: 125), or alternatively, as the dark side of European modernity (Mignolo, 2011). Mignolo, who further developed Quijano's arguments, has emphasized the importance of 'praxis' as the goal of decoloniality. This involves 'undoing' Eurocentric orders of knowledge, but also 'reconstituting forms of life that we'd like to preserve' (Mignolo and Walsh, 2018: 120).

Building on the 'coloniality of power' argument, Lugones turns her attention to gender, which is under-theorized in Quijano's model (2007). She shows how Eurocentric, patriarchal categories of gender were imposed on societies that had completely different understandings of gender, sex and sexuality before their colonization (Lugones, 2007: 196–8). Just as race was an invention that was imposed upon the colonized, so were patriarchal gender categories, which 'violently inferiorized colonized women' (Lugones, 2007: 201). This is illustrated in Oyéronké Oyěwùmí's study among the Yorùbá, who had no gender system in place before their colonization (1997). 'For females, colonization was a twofold process of racial inferiorization and gender subordination [...] The creation of "women" as a category was one of the very first accomplishments of the colonial state' (Oyěwùmí, 1997: 124). While the Yorùbá pre-colonial state organization was not gender determined, women – once created – were thereafter excluded from positions of power in the Eurocentric model of the state. One way through which coloniality survives today is through the persistence of the Eurocentric and patriarchal categories of race and gender.

There are clear parallels here with arguments in postcolonial theory. Said has argued that the claim to universality by European thought is a construction that is sustained through economic and political power globally (Said, 1978). The othering of Eastern cultures through the ideology of orientalism helped to legitimate colonial projects and the material domination of Western powers (Said, 1978). The consequence of orientalism is that its others are rendered mute and can only be understood through theories of the West. Spivak developed the notion of 'epistemic violence' to refer to the processes of othering when subaltern groups are rendered voiceless and can only be understood through colonial frameworks (Spivak, 2010). The ongoing imposition of Eurocentric knowledge and the parallel destruction, or invalidation, of 'ways of knowing that do not fit the dominant epistemological canon' are also understood as 'epistemicide' (Santos, 2016: 238).

Spivak is critical of the complicity of Western intellectuals and universities in the contemporary reproduction of colonial epistemic orders. This 'sanctioned ignorance' takes place through deliberate omissions or selective readings, which silence non-Europeans (Spivak, 1999). Sanctioned ignorance is common in fields such as international development, where people of the 'developing world' are understood through theories of the West (Escobar, 1995). There are many parallels between the study of humanitarian emergencies and international development – in both fields people in poorer countries have been constructed as a 'problem' waiting for solutions from the West in order to be lifted out of poverty. These solutions can be technical, economic and almost always entail a vision of European modernity. In Chapter 1 we will further unravel the relationship between humanitarianism, empire and capitalism.

Empires collapse, but they leave behind debris – and these ruins are durable and can be reactivated under different conditions, often in oblique ways (Stoler, 2016). Ann Stoler argues that these durable and hardened 'ruins' of colonial pasts produce contemporary 'imperial formations' evidenced in processes like displacement or refugee camps (Stoler, 2016: 56). It is important to clarify here that the emphasis is not on a single or dominant sovereign empire, but rather on the intangible, affective and protracted processes that suffuse people's lives and are passed on from generation to generation (Stoler, 2008). Crucially, these connections are often 'occluded' and thus mistakenly assumed to be new and distinct from empire (Stoler, 2008: 193–4). The generative approach of 'imperial formations' invites us to 'reassess what constitutes contemporary colonial relations' (Stoler, 2016: 341).

Understanding 'coloniality' as a diffused process explains why we still encounter it *within* postcolonial states, in the ways governments and bureaucrats often internalize practices of empire. This is a form of 'internal colonization' when local elites and intellectuals emulate the West and end up reproducing the colonial structures of oppression (Cusicanqui, 2012: 97). State bureaucracies continue to be organized according to the 'script of the colonizers' (Mignolo and Walsh, 2018: 122). The whole idea of the 'state system' with its 'bureaucracy' has always been integral to European modernity, and possibly one of colonialism's most enduring legacies. In his ethnography of bureaucracy in India, Akhil Gupta shows how certain practices of enumeration and inscription replicate those of the British colonizers (Gupta, 2012). Paraphrasing Bruno Latour, Gupta exclaims: 'we have never been postcolonial' (Gupta, 2012: 23). Bhabha has captured this ambivalent relationship between colonizers and colonized through the notion of 'colonial mimicry' (1994). The internalization of coloniality can be discerned in the way that governments in the global South internalize discourses about modernity and economic development, which are usually associated with institutions such as the World Bank. Such discourses are evident the building of infrastructural projects in Nigeria (Larkin, 2013) or in the championing of technological innovation in China (Lindtner, 2021).

A clear theme transpires from our discussion so far. Colonialism emerges as a tenacious structure well after the collapse of empires. This is evident in the persistent processes of othering, the epistemic violence of Eurocentric knowledge systems when they claim to be universal, and in formations of race and gender. Capitalism plays an important role here as it is entwined with colonialism, race and gender. Cedric Robinson's (2021) analysis of 'racial capitalism' reveals how capitalism developed along racialized lines, extracting value from marginalized and racialized people. The resurgence of interest in Robinson's work has also included a feminist lens that highlights how capitalism has depended on the gendered as well as racialized division of labour (Bhattacharyya, 2018).[10] These structures, these 'hardened ruins' of empire (Stoler, 2016), are internalized by postcolonial states, but also by individuals, which is why colonialism produces subjectivities, an argument compellingly made by Frantz Fanon (1961). There are parallels here with settler colonial studies and the analysis of colonialism as an ongoing 'structure' to explain how the legacies of settler colonialism persist and shape contemporary formations of race, gender and class (Glenn, 2015; Wolfe, 2006).

By 'technocolonialism' I acknowledge the enduring structures of domination in processes like displacement, migration, humanitarianism, and the development of science and technology.

Let's start with humanitarianism. Humanitarianism is deeply entangled with colonial histories. Humanitarianism emerged in the colonial expansion of the nineteenth and twentieth centuries (Lester and Dussart, 2014). Although contemporary humanitarianism is popularly understood in moral terms as the expression of 'a supposed natural humaneness' (Fassin, 2012) and the 'imperative to reduce suffering' (Calhoun, 2008), the structural asymmetry between donors, humanitarian officers and aid recipients reproduces the unequal social orders that shaped colonialism and empire. The emphasis on 'doing good' occludes the fact that aid, including international development, is part of a wider liberal agenda (Escobar, 1995) that primarily benefits countries in the global North. More broadly, humanitarianism reproduces relationships of inequity between Western 'saviours' and the suffering former colonial subjects, thus attesting to the tenacity of colonialism.

Migration and displacement are often the historical products of the aftermath of European colonialism (Bhambra, 2017; Hegde, 2016; Khiabany, 2016). Crucially, the colonial order is reproduced through the racial subjugation of migrants, which is why the so-called European migration and refugee crisis has been analysed through the coloniality of power (De Genova, 2016: 75). Refugee camps are another example of the tenacity of colonialism. Stoler (2016) identifies continuities between the nineteenth-century penal colonies and contemporary refugee camps or detention centres. Just like their nineteenth-century counterparts, refugees are 'abandoned people' (Mbembe, 2017: 3) yet closely monitored. Displaced people today have become 'the toxic refuse of our contemporary world' (Stoler, 2016: 337), taking the place of previous marginalized populations.

Technology and science are also part of colonial genealogies. Science was integral to the 'civilizing mission' of colonialism, a tool used to justify colonial rule (Fanon, 1959). Science was the prime tool used to mould colonial subjectivities (Anderson, 2006), for example through systems of classification. AI is part of larger genealogies of enumeration, measurement and classification that were originally developed by imperial powers to control colonial subjects (Appadurai, 1993). It is no coincidence that biometrics was first used in India as part of the British Empire's efforts to control and manage colonial subjects (Cole, 2001). That biometrics is still used today to manage and control othered bodies is evidence of the durability of colonial legacies (Madianou, 2019b). Infrastructures, too, have colonial legacies most evident in the infrastructure of the telegraph, on which the internet's infrastructure is built (Starosielski, 2015). The colonial genealogies of science are most evident in the way refugee camps are referred to as laboratories for experimentation, echoing the medical and pharmaceutical experiments of the previous century (Petryna, 2009).

My use of the term colonialism does not refer to a radically new phase of colonialism, neither does it suggest a historical prism through which to understand the present. My argument is that colonialism has never gone away. Empires collapsed, but their legacies and logics have survived and permeate processes such as humanitarianism and its institutions. They also permeate technology, which in turn reworks and revitalizes colonial relations. But these relations are part of the present and they have been around for a long time.

Likewise, I do not use colonialism as a metaphor (Tuck and Yang, 2012). It would be hugely problematic to do so, given the violence associated with colonialism. To use colonialism as a metaphor is to depoliticize the term and the phenomena it describes.

I argue that we need colonialism as a framework in order to explain why technological experiments take place in refugee camps, typically in the global South. We need colonialism to understand why there is no meaningful consent in technological experiments in refugee camps, where refusing to submit your biometric data amounts to refusing to receive aid and shelter – when there are no other alternatives for survival. We need colonialism to understand why chatbots, trained in the English language, impose Eurocentric ideas about illness or economic rationality on people living in precarity, who have different understandings and priorities. We need colonialism to understand the epistemic violence of biometric technologies, which impose prototypical 'whiteness' on othered bodies (Browne, 2015) and therefore discriminate. Technological interventions materialize and entrench existing forms of discrimination and turn them into scientific facts. But I am rushing. Before I develop technocolonialism – we need to make sense of digital technologies, which I analyse as a process of infrastructuring.

Infrastructures

Infrastructure is a productive framework for making sense of emergency contexts, while emergencies themselves are revelatory of the workings and politics of infrastructures (Finn, 2018; Graham, 2010). Moments of failure make infrastructures visible and reveal the power dynamics that sustain them and are sustained by them. Disasters involve the breakdown of infrastructures, while disaster recovery concerns the attempt to repair and rebuild them. Such moments of disruption can reveal the contrasting imaginaries of technology. In the days after Typhoon Haiyan made landfall, Google staff were busy coding their 'person finder' application to help survivors locate their missing family members. In Tacloban, where internet and all communications infrastructure had been completely destroyed and remained cut off for weeks, people developed their own alternative infrastructure using pieces of carton, or paper plates on which they wrote the names of the missing and which they stuck on the walls of the buildings that survived the storm surge.

It was my research fieldwork that led me to the study of infrastructures and not the other way around. During my fieldwork I became increasingly aware that biometric data, digital identity databases, feedback data, drone and satellite data began to underpin humanitarian operations. Biometric data in particular were becoming ubiquitous in refugee camps and supported the delivery of basic services such as cash assistance. Many of these systems are becoming increasingly interoperable with state bureaucracies and with commercial platforms, such as messaging apps. The sheer size of commercial platforms such as Facebook and WhatsApp means that they, too, become de facto infrastructures (Plantin and Punathambekar, 2019). This is especially the case among low-income communities in countries in the global South where, in the absence of public infrastructures, these new private systems provide the first experience of networked communication. Because Meta has a quasi-monopoly over technologically mediated communication, its platforms (Facebook or WhatsApp) acquire features of infrastructures such as scale and indispensability (Plantin and Punathambekar, 2019). As I began to follow the trails of data among humanitarian workers, donors, government departments, private companies and data subjects it became clear I needed an expansive framework that could encompass the networks, data flows, technologies, platforms, devices and the relations among them. Crucially, an infrastructural approach allowed me to map the power relations between the various actors and the constituent elements of the system. This was very pertinent for the study of humanitarianism, which is suffused in relations of power.

An infrastructural approach offers an analytical lens to combine the study of the networks and flows of data, as well as the constituent technologies (AI, biometrics, blockchain) and platforms (messaging apps and social media platforms). By combining these into the book's analytical framework, I contribute to the convergence between infrastructure and platform approaches that bring together the fields of science and technology studies, information studies and critical algorithms studies (Plantin et al., 2018). I also draw from the study of infrastructures in anthropology and in particular the work of Brian Larkin (2013).

Infrastructures are 'built networks that facilitate the flow of goods, people or ideas and allow for their exchange over space' (Larkin, 2013: 328). In that sense, infrastructures encompass but extend beyond the notion of 'substrate', the idea of a system such as a railroad on which rail cars run (Star and Ruhleder, 1996: 116). Humanitarian infrastructures include the substrates as well as the data, technologies, ideas about technology and affects that flow through them. Humanitarian infrastructures encompass humanitarian workers, host governments, donors (which largely includes Western governments), private companies, volunteers, digital developers and crisis-affected communities and the relations between them.

My understanding draws on, but also expands on Star and Ruhleder's widely used, but rather technical definition of infrastructures as 'embedded', 'invisible', 'shaped by pre-existing installed bases', 'learned as part of community membership' and 'linked to established conventions of practice' (1996: 113). Humanitarian infrastructures are 'embedded' in existing systems such as the web, or social media, but also the humanitarian system itself. They build on, are shaped and constrained by pre-existing 'installed bases' such as undersea cables, cloud computing, blockchain systems and satellites among others. Once established, humanitarian infrastructures become invisible or 'transparent' as happens with biometric or digital identity systems, which move to the background and mostly become visible at moments of breakdown or errors. Humanitarian infrastructures are 'learned' as part of people's involvement in the humanitarian space and linked to established 'conventions of practice', such as feedback mechanisms or cash assistance (see Star and Ruhleder, 1996: 113). Because the field of humanitarianism is fragmented and governed by different logics, it is important to emphasize here that infrastructures, and the 'conventions of practice' to which Star and Ruhleder (1996) refer, are inherently contested. For example, biometric systems are viewed and experienced differently by refugees, humanitarian workers, governments, and the private sector. The experience of biometrics is radically different depending on whether someone is a refugee or a humanitarian worker. Because humanitarian infrastructures are imposed top-down, it is important to take into account the power dynamics involved within the context of the historical genealogies of communication infrastructures. In order to capture the relational, dynamic, contested and open-ended character of infrastructures as processes, I draw on the term 'infrastructuring', which has been proposed to emphasize those traits (Karasti and Blomberg, 2018).

Just like science and technology, the development of infrastructures is steeped in colonial relations (Cowen, 2020) and imperial material domination (Aouragh and Chakravartty, 2016). Infrastructures are historical, in the sense that they build onto existing infrastructures. The archetypical example is the telegraph, which is a product of the imperial expansion of the nineteenth century (Supp-Montgomerie, 2021). The telegraph cable systems mapped onto the geography of the British Empire (Starosielski, 2015) and this infrastructure provided much of the footprint for later infrastructural iterations such as the internet. Not only was the telegraph key for the administration of the colonies of the British Empire, it was also the result of a massive programme of extraction of natural resources – mostly gutta-percha, a type of rubber derived from trees in South-East Asia, which was necessary for insulating the vast network of underwater cables – and human labour from the then colonies (Tully, 2009). Seuferling and Leurs (2021) have historicized the infrastructures of refugee governance by documenting their origins in the immediate aftermath of the Second World War. A historical approach to infrastructure allows us to move beyond the preoccupation with the technical aspect of the system and to examine the genealogies of infrastructures and their contemporary iterations.

Infrastructures are assumed to be boring, invisible or banal, but that should not distract from the fact that they are inherently political. Infrastructures are often assumed to exemplify the 'Enlightenment dream of social control' (Graham, 2010: 4). This is a teleological – and Eurocentric – understanding of technology and infrastructure where each iteration of the technology is assumed to be an improvement compared to earlier versions (Kember and Zylinska, 2012). The most perfect infrastructure is the one that disappears into the background and becomes invisible. Yet, this description of infrastructure is only a fantasy or wishful thinking, especially in poorer parts of the world where infrastructures can be unreliable, fragmented, problematic and therefore very visible. Infrastructures are not neutral. They are asymmetrically distributed, resulting in vastly differentiated experiences depending on where one lives in the world. The political character of infrastructures is also evident in their capacity to organize the world based on classifications.

Infrastructures depend on classifications that are inherently subjective (Bowker and Star, 1999). For instance, biometric systems depend on classifications regarding the physiological characteristics of the human body. As we'll explore in detail in Chapter 2, these classifications are not objective, but reflect racist and gendered biases with colonial genealogies (Browne, 2015; Magnet, 2011; Buolamwini and Gebru, 2018). The invisibility of infrastructural classifications does not detract from their power. Quite the contrary, it places more urgency on recognizing infrastructures as sites of significant political power.

Also hidden is the material stuff of infrastructure: the cables, pipes and the virtual 'cloud' are all kept out of sight. Infrastructures are 'camouflaged by design', a process that is captured by the term 'infrastructural concealment' (Parks, 2012). Here we observe a parallel with theories of technological mediation (Bolter and Grusin, 2000; Eisenlohr, 2011; Meyer, 2013). Mediation tends to erase its own work, in other words, to become absorbed into whatever it mediates. John Durham Peters is fascinated with infrastructure's tendency to 'recede into the background' (2015: 35). His approach of 'infrastructuralism' focuses on understanding the work below the surface: the job of ordering, of organizing and orienting, of arranging relationships between people and between things (Peters, 2015: 37). Revealing the workings of infrastructure is key for mapping the power relations therein.

Humanitarian infrastructures are typically imposed from the top down. Of course, people affected by crises use social media and 'public information infrastructures' (Finn, 2018). But bespoke humanitarian systems are distinct in that they are imposed by relief, government or private organizations. In some cases, such as biometric enrolments, engagement with the infrastructure is conditional for receiving relief. Unless a refugee is biometrically registered, they will not be counted as a refugee and therefore they will not receive aid (or any associated rights that come with refugee status). Recognizing the coercive nature of humanitarian infrastructure is an important starting point of my analysis.

At the same time, I acknowledge that even within these very asymmetrical relations, people express their agency through acts of appropriation or resistance. Infrastructures are both tools of control and resistance (Cowen, 2020). Crisis-affected people do not just engage with humanitarian infrastructures, but they have access to a range of communication channels – even though these may be restricted due to disaster or lack of resources. It is important to recognize that even in asymmetrical settings infrastructures need to be understood relationally (Larkin, 2013).

Infrastructures are material and immaterial things – but also the relations between things. For example, a biometric system makes tangible the relationship between data subjects (refugees), humanitarian workers, host government officials, donor governments such as the USA and the private contractors who usually run the system. Tracing the flow of biometric data, which is enabled by infrastructures, allowed me to discern the power geometries in the humanitarian space. Infrastructures are not just technical structures, but also 'structures of feeling' produced through breakdown or exclusion. The earlier example regarding the use of Skype for the processing of asylum applications in Greece is a case in point. In the Skype example, the encounter with the infrastructure engendered frustration, despondency, angst and anger for those who were unable to get the system to offer them an appointment. By providing the 'ambient environment of everyday life', infrastructures shape subjectivities (Larkin, 2013: 333).

At the same time, infrastructures are shaped differently depending on who uses them. Refugees and humanitarians have a wildly contrasting experience of biometrics – with refugee bodies always being the ones measured, and humanitarians conducting or processing those measurements in order to release assistance. To understand the divergent experiences of infrastructures, which is essential for understanding the power relationships at stake, I take an ethnographic approach. Starting with the people affected by Typhoon Haiyan and their encounters with humanitarian infrastructure, I then expand to the practices of humanitarians, government officials, entrepreneurs, technology companies among others. My analysis will trace infrastructures both at the level of everyday life, but also at the organizational and inter-organizational level, along the micro and meso levels suggested by Edwards (2002) (see 'Note on Methods' for further discussion).

Infrastructure is an expansive concept, which is part of its appeal. Infrastructures contain a range of technologies, which also need to be understood in their own right. By emphasizing infrastructure, we should not gloss over the particularities of the different technologies and platforms and how they themselves shape humanitarian practices and are shaped by them. I discuss biometrics and blockchain in Chapter 2, mobile phones, texting and messaging apps in Chapter 3, AI and chatbots in Chapter 4 and AI and automated decision making in Chapter 5.[11] But because all of these technologies, platforms, devices and associated practices co-exist and intersect within a larger system, we need an all-encompassing notion – infrastructure – to make sense of them as a sociotechnical assemblage.

The poetics of infrastructure and the imaginaries of 'technology for good'

Infrastructures encompass ideas about technology, what Larkin calls the 'poetics' of infrastructure. Infrastructures acquire powerful meanings, 'almost fetish-like aspects that sometimes can be wholly autonomous from their technical function' (Larkin, 2013: 329). One such example is the teleological understanding of AI as synonymous with the future and as a solution for the challenges of humanitarianism (among other things). This echoes Taylor's notion of social imaginaries as 'what enables, through making sense of it, the practices of a society' (Taylor, 2002: 91). Social imaginaries include the discourses and ideologies about digital technologies and practices of computation. For example, popular phrases like the 'digital revolution' – and, more recently, the 'big data' or 'AI revolution' – do important work in framing expectations regarding the potential of technologies to catalyse change.

The social imaginary surrounding the term 'technology for social good' deserves our attention. Digital innovation in the humanitarian sector is often described as part of the 'technology for good' initiatives. There are many iterations of this phrase: 'technology for good', 'AI for good', 'Blockchain for good', 'Virtual Reality (VR) for good' and 'infrastructures for good'. The 'for-good' element of the phrase, makes a moral claim about the purpose of technology, while occluding the underlying power relations. By foregrounding technology as 'good', 'technology for social good' forecloses critical questions regarding the asymmetries of humanitarianism. Likewise, the emphasis on 'good' obscures the way technology companies often use 'good' projects to entrench their position in public life. Part of the problem is that the term 'technology for good' assumes what the 'social good' is, when the term is fundamentally contested. What is good for one group, may not be good for others. Who decides what the good is is crucial – and this is a matter of power, not ethics.

If some consensus exists about the definition of good, it revolves around the orientation towards the UN's sustainable development goals (SDGs), which are the focus of the annual 'AI for good' Summit, which is hosted by the UN and brings together the 'AI innovators' or 'problem solvers' with the world's 'problem owners' to refer to the terms used by the Summit organizers.[12] By closely aligning to the framework of international development, AI for good inherits some of the criticisms regarding development and humanitarianism discussed earlier, namely the reproduction of asymmetrical relationships between the global North and South, which is reflected in the language about 'problem owners' and 'problem solvers' that the AI Summit foregrounds. By taking a top-down approach that presupposes what the good is, AI for good projects revive some of the long-standing criticisms about development and humanitarianism as preserving Eurocentric systems of knowledge and, ultimately, the coloniality of power (Quijano, 2000). The book will unravel these assumptions behind 'technology for good'. Instead, the book asks: what animates the adoption of digital technologies and infrastructures in the humanitarian sector? What are the consequences of the digitization of humanitarian operations? And how do people affected by crises appropriate these technologies and infrastructures?

Technocolonialism

We can discern how colonialism and technological infrastructures combine to produce what I term technocolonialism. Technocolonialism reworks and revitalizes colonial genealogies through processes of extraction, coloniality, control and discrimination. Colonial structures are maintained through extracting value from the data of refugees and other crisis-affected people; and through extracting value from experimentation with new technologies in fragile situations for the benefit of stakeholders, including private companies. Technocolonialism involves the imposition of Eurocentric concepts and systems onto local cultures. This takes place through the introduction of Eurocentric ideas such as accountability via digital interventions such as feedback platforms. Because it relies on infrastructural classifications that have known biases in terms of race, gender, ethnicity and ability (Benjamin, 2019; Bowker and Star, 1999; Noble, 2018 among others), techncolonialism materializes and entrenches existing forms of discrimination. Although digital technologies are championed as means of empowerment, they are ultimately vessels of containment and instruments of control. Digital technologies are technologies of distancing. Because they involve multiple actors and complex supply chains, they dehumanize suffering and erase accountability. The context of emergencies justifies several of the above practices.

It is already clear that technocolonialism, is about violence. Not just symbolic or epistemic, but also physical violence as was evident in the Rohingya example discussed earlier. In fact, the potential harms associated with the Rohingya biometric registrations continue and extend into the future. In 2019 it was revealed that the Rohingya biometric records were indeed shared with the Myanmar government potentially risking further persecutions, this time supported with biometric data (Human Rights Watch, 2021). This is a clear example of function creep, which refers to the situation when data obtained for one purpose end up being used for something completely different.

Technocolonialism ultimately is understood as a form of structural violence which is enacted through the convergence of humanitarian, state and market forces. The term structural violence was developed by Paul Farmer, drawing on earlier work by John Galtung (1969), to refer to a host of violations of human dignity: 'extreme and relative poverty, social inequalities relating from racism to gender inequality, and the more spectacular forms of violence that are [...] human rights abuses' (2005: 8). Structural violence is linked to structural inequities and processes of exclusion and marginalization, which affect whole groups on the basis of their gender, race, class or where they live in the world. In that sense structural violence is systematic, but also experienced indirectly (Farmer, 2004: 307). A key feature of structural violence is its diffused nature, experienced through everyday humiliations and cruelties. Akhil Gupta in his ethnography of state bureaucracy in India observes how the state enacts structural violence through everyday acts of corruption, inscription and governmentality (2012).

Technocolonialism shifts the attention to the constitutive role of data and digital practices in entrenching inequities not only between crisis-affected people and humanitarian agencies, but also in the global context. I argue that the infrastructuring of humanitarian operations produces new forms of harm, discrimination and structural violence. We need a new term and theoretical framework to understand the constitutive role of infrastructures and technologies in this process. Neocolonialism, which refers to the control and exploitation of former colonies through economic policies and other indirect means, does not offer a framework to understand the critical role of digital infrastructures and technological mediation. This is why I propose a new term, technocolonialism.

The violence of technocolonialism is captured in Zach Blas' artwork. 'Face Cages', which is reproduced on the cover of the the paperback edition of this book, offers a powerful visual entry point to the topics explored here. 'Face Cages' dramatizes the abstract violence of the biometric diagram and, by extension, technocolonialism. In Zach Blas' work, the diagram, a supposedly perfect identification technology, transforms into a cage which imposes itself onto the human body. According to Blas, these cages 'exaggerate and perform the irreconcilability of the biometric diagram with the materiality of the human face itself – and the violence that occurs when the two are forced to coincide'.[13]

From the discussion so far, we can discern the different *dramatis personae* involved in technocolonialism. Humanitarian organizations (whether intergovernmental, like UN agencies, or large international non-governmental such as Oxfam and smaller NGOs), host governments, donor governments, private companies including technology companies and start-ups, digital developers, volunteers and, of course, communities affected by crises. It is important to recognize that technocolonialism is not just about digital capitalism. While we are witnessing an increasing involvement of private companies in humanitarian operations, the role of national governments is crucial. States are the largest funders of humanitarian organizations and contribute to vital decision making in the sector such as which emergencies will receive attention and which will not. States push for the implementation of digital infrastructures such as biometric systems in line with their securitization agendas – something we'll discuss in detail in Chapter 1. Further, governments are directly involved as hosts to refugee populations and when managing emergencies in their own territory. To understand technocolonialism, we must acknowledge that states, private companies, humanitarian organizations and digital infrastructures are equally implicated in reworking colonial genealogies. Chapter 1 will map the competing logics that animate these actors and explain why, despite their very different motivations, they all prioritize digital innovation. Recognizing the equal involvement of governments, capitalist forces, humanitarian organizations and the workings of infrastructures themselves is one of the fundamental ways in which technocolonialism differs from parallel terms such 'data colonialism' (Couldry and Mejias, 2019; Thatcher, O'Sullivan and Mahmoudi, 2017) or 'digital colonialism' (Kwet, 2019; Schneider, 2022), which emphasize the role of capitalism.

Technocolonialism is also distinctive in its approach, which foregrounds the perspective of people in humanitarian settings. Their voice, often absent in discussions about data, AI and technological infrastructures, is essential for understanding the harms of these systems, but also for understanding how technocolonial projects are contested and resisted. The postcolonial and decolonial approaches that inform my thinking about technocolonialism require that people living in precarity speak about their experiences in their own words. My initial ideas about technocolonialism began through the ethnography with communities in the aftermath of Typhoon Haiyan. While my later research offered a multi-optics approach (combining the perspective of humanitarian officers, government officials, designers, entrepreneurs and donors among others), the ethnography with the people of Tacloban and Sabay remains the foundation of my understanding of these processes.

Outline of the book

Chapter 1 identifies the key logics that explain the push for digital technologies in humanitarian operations. These logics explain the structural conditions for the emergence of technocolonialism and, in so doing, inform the analytical framework of the book as a whole. In order to make sense of the contemporary developments, the chapter will begin by providing some historical context for the development of humanitarianism. In the first part of the chapter, I trace this history by focusing on key aspects: the relationship between humanitarianism and empire, the relationship between humanitarianism and capitalism and the bureaucratization of humanitarianism.

The logic of humanitarian accountability is behind the assumption that the interactive nature of digital technologies will facilitate the participation of affected communities in their own recovery and thus improve the accountability of humanitarian operations. The 'logic of accountability' legitimates digital developments within the sector. For example, 'digital identity' initiatives, which use technology to identify and verify people, are framed as contributing to the recognition and empowerment of refugees (UNHCR, 2018). The 'logic of audit' stems from the constant demand for metrics which humanitarian organizations must submit to donors in order to secure more funding. Given the huge growth of the humanitarian sector, with the global aid economy estimated at over US $41.3 billion,[14] there is an acute pressure for audit. Digital technologies generate metrics and are associated with robust audit trails and efficiencies. The logic of capitalism explains the dynamic entry of the private sector in the humanitarian space through the now ubiquitous private–public partnerships. The idea that there can be technological fixes to complex problems exemplifies the logic of technological solutionism. The logic of securitization concerns primarily the role of the state and the way it uses technologies to make populations legible and protect borders. This is particularly relevant to the response to refugee issues and the use of biometric technologies. As always, structures of control produce contestation. The chapter identifies a sixth logic – the logic of resistance, which refers to the extent to which people challenge the infrastructures of technocolonialism. These are analytical distinctions. In practice these logics intersect and produce the phenomenon of technocolonialism.

We begin the exploration of digital humanitarian infrastructures with the introduction of biometrics in humanitarian settings. **Chapter 2** begins by tracing the genealogies of biometrics back to the nineteenth century practices of fingerprinting in the British Empire. The historical perspective reveals continuities with the contemporary uses of biometrics in humanitarian settings. In both cases, biometric registrations were introduced in very asymmetrical contexts to manage and control populations. Despite technological innovation, biometric methods continue to reproduce problematic classifications regarding race, gender, class, age and disability. Biometric technologies 'privilege whiteness' (Browne, 2015) with significantly higher margins of error when measuring, or verifying 'othered bodies', whether in terms of race, ethnicity, gender, class, disability or age. Rather than being the perfect identification technologies as they are often referred to, biometric systems codify existing forms of discrimination (Browne, 2015; Magnet, 2011).

The introduction of biometrics in refugee camps is a form of epistemic violence which reduces identity to a transactional category while stripping people of their agency to define themselves. Concerns regarding privacy, safeguarding and function creep add to the potential harms which amount to structural violence. The lack of meaningful consent in refugee biometric registrations further compounds some of the above risks. While it is theoretically possible for a refugee to refuse biometric data collection, this is not an option for most refugees as that would amount to refusing aid when no other livelihood options are available. Ultimately, digital identity practices reconfirm the hierarchy between aid providers and refugees – and in so doing reaffirm that, structurally, contemporary versions of humanitarianism are not dissimilar to their colonial counterparts.

Chapter 3 turns our attention to practices of accountability and how they are ultimately repurposed as processes of extraction. Extraction does not just refer to material resources, but also to data (Mezzadra and Neilson, 2019). Affected communities, through their data practices, produce value which is extracted for the benefit of humanitarian agencies and other stakeholders. This is evident in the case of feedback practices which are mandatory in aid operations and increasingly digitized through instant messaging platforms (such as WhatsApp) or chatbots. Although feedback apps appear to be addressing the demands for humanitarian reform and greater accountability, in practice, they serve the logic of audit by satisfying donors' demand for impact data. Data are extracted to legitimate humanitarian projects, but are rarely used to actually improve humanitarian operations. This chapter draws on the Typhoon Haiyan fieldwork and traces the flow of feedback data through humanitarian infrastructures. Rather than being fed back to local communities in order to improve the delivery of aid, feedback data flow from affected areas to the headquarters of humanitarian agencies and then to donors. Digital affordances streamline feedback and package it into databases which provide tangible evidence about the impact of relief programmes. The datafication of feedback facilitates the turning of accountability into audit. That these shifts take place in the name of greater 'accountability to affected people' means that the continuing power imbalance is cloaked under the promise of digital efficiency and transparency. This is accountability in appearance, which disenfranchises humanitarian subjects. Chapter 3 also observes the coloniality of feedback mechanisms as essentially Eurocentric practices that do not resonate with the priorities of local communities. Crucially, even when feedback mechanisms fail, they succeed in legitimating humanitarian projects.

Chapter 4 continues to unravel the theme of extraction through the proliferation of technological pilots which are becoming ubiquitous in humanitarian settings. The chapter begins with a historical account of nineteenth and twentieth-century experimentation which explains why people in the global South are readily available as experiment subjects. In the contemporary context, refugee camps are treated like 'laboratories' where new technologies are piloted by private companies often in partnerships with humanitarian organizations. The hype generated by technological pilots (such as 'chatbots', 'blockchain cash assistance' or other 'AI for social good' applications) ultimately benefits the technology companies which cultivate the perfect branding opportunities for their products.

By seeking problems for solutions, technological experiments are at odds with the humanitarian imperative 'do no harm'. Moreover, 'artificial intelligence for social good' projects reproduce power asymmetries by asserting Eurocentric values. Experimentation follows a North–South trajectory while projects often lack linguistic or cultural sensitivity and impose instead imported categories on local contexts. Because technologies are cast as enchanted objects, the power asymmetries of humanitarianism are heightened.

Finally, Chapter 4 observes a new type of experimentation, which I term surreptitious experimentation, which refers to technological trials that take place outside the laboratory, without clear boundaries or processes of accountability. In so doing surreptitious experimentation further compounds the asymmetries within the humanitarian space.

Chapter 5 charts the emergence of the 'humanitarian machine', a symbolic and material infrastructure that brings together the different stakeholders involved in the humanitarian field. The humanitarian machine is the outcome of the increasing bureaucratization of humanitarianism and the parallel infrastructuring that takes place through processes of digitization. By infrastructuring I refer to the fact that digital systems and practices of computation underpin the humanitarian space as a whole. Biometrics is a good exemplar of this. Biometric databases increasingly underpin other systems, while networks become – or aim to become – interoperable.

Automation and computation accentuate the known shortcomings of bureaucratic systems. Refugees flee war and violence, only to be met with a nonhuman machine that bureaucratizes their experience and engenders dispossession. Automated decision making excludes and discriminates. The rigidity of the machine coupled with data sharing agreements with governments produces errors with catastrophic consequences for minoritized people. Because digital systems are infrastructural, they rely on invisible and opaque processes, which means that any errors are hard to redress. Harms also occur at the everyday encounters with the machine: with the chatbot that doesn't include one's problem as part of the drop down list, or the technology that doesn't work. Further, the labyrinthine system of supply chains and automated decision making obliterates accountability to affected communities.

Processes of automation do not only accentuate the power asymmetries between humanitarian agencies and affected communities: they also contain the urgency of humanitarian crises by preventing them from erupting. The machine prevents us from addressing the humanity of the crisis, the machine keeps the crisis out of sight.

Chapter 6 returns to the micro level of everyday life and observes the expressions of resistance in asymmetrical settings. The example of the Rohingya with which we began the book is a reminder that despite power asymmetries, technocolonialism is met with resistance. Even though the strike was quashed by police force, it confirms the contested nature of technocolonialism. Most contestation in humanitarian settings is not overt. Chapter 6 develops the notion of 'mundane resistance' in order to make sense of the micropolitics of everyday life.

The literature on protest – which is largely Anglocentric – understands resistance through expressions of activism, outright dissent or revolution. However, the cost of rebellion is prohibitive in many parts of the world. In order to make sense of resistance in asymmetrical settings I draw on writers from the Black radical tradition, like Orlando Patterson and Cedric Robinson who observed that in very asymmetrical settings such as slavery, resistance took passive forms, including deliberate evasion, refusal to go to work, or satire (Patterson, 2022; Robinson, 2021). Mundane resistance is a term that captures the ordinary and latent nature of resistance through small acts like non-participation or the oppositional uses of infrastructures. These may be small acts of resistance, 'in a different voice' (Gilligan, 1982), but they are not necessarily passive. The chapter also identifies other forms of resistance from open activism to digital witnessing and storytelling.

The field of media research was once described as series of oscillations between approaches that either favoured powerful media, or powerful audiences. In this juncture of internet research, the pendulum is firmly on the side of technological and data power. I argue that such stark choices can be false: technologies are powerful and rework colonial genealogies; at the same time, it's important to acknowledge people's agency, however limited, to contest structures of oppression and to articulate their own identities. It is important to stress the limits of everyday resistance in asymmetrical settings such as those encountered in the humanitarian sector. Mundane resistance does not equal decolonization, which requires a radical restructuring of power relations (Smith, 2021). Still, ordinary resistance needs to be acknowledged for it may contain the seeds of the decolonial struggle to come.

The **Conclusion** draws out the implications of the pervasive infrastructures that underpin the humanitarian space. I argue that we are witnessing a structural transformation of humanitarianism and its relationship with states and the private sector. As the boundaries between humanitarian organizations, nation-states and private companies become porous, there are implications for governance, accountability and even the identity of relief organizations which have always tried to distance themselves from politics. The infrastructuring of the humanitarian sector engenders a new form of structural violence, which I term 'infrastructural violence', which is normalized and legitimated by the practices of digital consent and feedback mechanisms and by the overall enchantment with technology.

A note on language and terminology

Language matters, especially when studying power relations. Language can reproduce marginalization and structural violence. Ngũgĩ wa Thiong'o (1986) reminds us that language is a key tool of othering, which is a form of epistemic violence (Spivak, 2010). The term 'aid' reflects a hierarchy of countries, where those with resources support the ones in need. Similarly, 'development' assumes a hierarchy, where some countries are more developed than others, based on a universalist idea of progress. The terms 'beneficiaries' and 'donors' also reflect these inequities and, like the previous ones, conceal the factors that contribute to global inequalities. To the extent that I describe these phenomena, I still use the term 'aid' (as in the 'aid industry' or 'aid sector'), but embed it in a critical discussion of humanitarianism (Chapter 1). When I refer to official policy I place terms in inverted commas. I avoid terms like 'beneficiaries' or 'humanitarian subjects' and refer to people instead.

In the book, I use the terms 'global South' and 'global North' as well as 'majority/minority world' to refer to the inequalities between countries. These terms have been proposed to replace deeply problematic terms such as 'developing nations'/'developed world', which imply a progression towards a Eurocentric modernity, while concealing the histories of extraction and violence under colonialism and capitalism. The terms global South/global North do not refer to geographic locations, but rather to the relations of extraction, oppression and coloniality between different parts of the world (Medrado and Rega, 2023). One advantage of the term global South is that it evokes emerging political solidarities (Medrado and Rega, 2023) and forms of knowledge production (e.g., 'epistemologies of the South', Santos, 2016).

The term 'majority world' refers to the countries where most of the world's population (the 'global majority') lives. 'Minority world' refers to the countries where a smaller proportion of the world's population (the 'global minority') lives. The majority world is here understood as synonymous with the global South (same as with minority world and global North). Majority world recentres the people of the global South as a majority. No term is perfect, of course. All broad categories risk glossing over differences. Not all countries in the majority world/global South experience inequalities the same way. Additionally, there are significant differences among groups within countries in the majority world, the same way we may find pockets of the 'South' in the global North (Milan and Treré, 2019).

Finally, a note on pronouns. I use the first-person singular ('I') when I describe my own understanding, interpretation or experience. I also use 'we' when referring to the collaborative fieldwork during the aftermath of Typhoon Haiyan in the Philippines ('the Haiyan project'), which informs Chapters 3 and 6. In the Note on Methods, I include a reflection on my own position in the research and writing process, as well as a discussion of the research process and data collected.

1

The Logics of Digital Humanitarianism

From the biometric registration of refugees and the use of blockchain for cash assistance, to the proliferation of chatbots for information dissemination and feedback, digital technologies and computation have become ubiquitous in humanitarian operations. These developments are often summarized in the term 'digital humanitarianism', which is a shorthand to refer to the uses of data, digital innovation and artificial intelligence in humanitarian emergencies. This chapter will trace the logics that explain the enthusiastic adoption of digital technologies and AI for humanitarian operations. In order to do so, we will begin with a brief historical account of humanitarianism which will allow us to examine its relationship with empire and colonialism. I will then unravel the relationship between humanitarianism and capitalism, before examining the bureaucratization of humanitarianism and the arrival of digital humanitarianism.

We refer to humanitarianism as if it were one thing, but in fact it is several. Humanitarianism is a historical phenomenon with roots in the imperial expansion of the nineteenth and twentieth centuries (Lester and Dussart, 2014), an industry, a profession, an ideology, a moral discourse and concern for the suffering of distant others (Calhoun, 2008; Chouliaraki, 2013; Fassin, 2012; Ticktin, 2011). Alleviating the suffering of distant others involves logistics that underpin complex operations and encompasses multitude of actors (Barnett, 2011). Another way in which humanitarianism is conceptualized is as a space, or a field, in which different actors operate.

In the Introduction we have already discerned the different actors that play a role in digital humanitarianism developments. Apart from humanitarian organizations, we observe the involvement of nation states, the private sector and, of course, the communities affected by war, conflict, famine, epidemics or disaster. Humanitarian organizations include agencies of intergovernmental bodies such as the UN, but also international non-governmental organizations (INGOs). Nation states are involved as donor governments, as hosts for refugees, or as recipients of aid. Commercial companies are involved through public–private partnerships with aid organizations, as vendors (for example, in the supply of infrastructure and service delivery of biometric enrolments) and, increasingly, as humanitarian organizations in their own right. Examples here include various philanthropic foundations such as the Bill and Melissa Gates Foundation. Technology companies are increasingly present in the humanitarian space, including start-ups and private entrepreneurs, who are involved in the design, management and implementation of digital platforms. Local communities include aid recipients, but also humanitarian workers who are employed by UN agencies, INGOs and local relief agencies. Other players include volunteers or grassroots humanitarians (Clayton, 2020) as well as digital volunteers, often referred to as 'digital humanitarians' (Meier, 2015), who become involved remotely through technology events such as hackathons – the collaborative events that aim to address social issues through the development of new software programmes – or mapathons – the collaborative online mapping using satellite imagery and open-source data for the use of humanitarian organizations.

Even from this brief overview it is clear that the humanitarian space is fragmented in the sense that the above constituents do not necessarily share the same principles, motivations, goals or understandings of what constitutes an emergency and how to respond to it. Some of these actors have entered the field recently. The private sector is more involved in contemporary humanitarian action than ever before. The chapter's primary aim is to identify the key logics that underpin the contemporary transformations of humanitarianism and in particular the push for digitization and datafication. In order to make sense of the contemporary context, the chapter will begin by providing some historical background for the development of humanitarianism. In the first part of the chapter, I trace this history by focusing on key aspects: the relationship between humanitarianism and empire, the relationship between humanitarianism and capitalism and the bureaucratization of humanitarianism. My analysis is inevitably selective. To cover the history of humanitarianism would require a whole volume and there are excellent books for that purpose.[1] The aim of the first part of the chapter is to provide enough context in order to develop the six competing logics which explain the enthusiasm for digital innovation in the humanitarian space. The six logics, which are the focus of the second part of the chapter, provide the analytical framework for the rest of the book.

A brief history of humanitarianism: Empire, capitalism and bureaucracy

The origin story of humanitarianism begins with Dunant, the Swiss businessman who, on accidentally stumbling upon the carnage of the battle of Solferino between the French and Austro-Hungarian armies in 1859, started a grassroots campaign that led to the establishment of the International Committee of the Red Cross (ICRC) and the Geneva Conventions. This story has an almost mythical function for the field of humanitarianism. Although compassion towards others has existed throughout the history of humanity, the nineteenth century marked a significant transformation: the institutionalization of compassion towards distant others. Despite the many definitions and approaches to humanitarianism, there is agreement that it refers to the organized forms of compassion and assistance towards suffering distant others (Barnett, 2011; Calhoun, 2008). The establishment of the ICRC as the first international humanitarian organization marked the beginning of the formal organization and *institutionalization* of compassion. The focus on *distant others* is the result of the particular historical period of imperialism. I will explore the two aspects of humanitarianism in turn.

The institutionalization of compassion

The institutionalization of compassion through organizations such as the ICRC came with a set of rules regarding its remit and practice. These are contested of course, but I will try to summarize the key points here. One strand of humanitarianism focuses on the provision of immediate relief during emergencies such as conflicts, disasters of epidemics. This is exemplified in the work of ICRC and Médecins Sans Frontiers (MSF) and the version of emergency humanitarianism which they represent. Other organizations have adopted a wider remit combining emergency work (e.g., saving lives) with more long-term assistance that resembles more traditional development work. Overall, we observe an increasing blurring of the boundaries between humanitarian assistance and international development (Krause, 2014: 108–9). Many of the largest NGOs, such as Oxfam, specialize in both, because they recognize that recovery extends beyond the immediate aftermath of an emergency. These divisions in the field are still deeply felt with the ICRC and MSF defending the emergency strand as a pure and apolitical form of humanitarianism, and the blended or 'alchemical' camp criticized for dabbling into the sphere of politics (Barnett, 2011).

The preoccupation of humanitarianism with the idea of emergency deserves some attention. According to the Oxford dictionary, 'emergency refers to a sudden serious and dangerous event or situation which needs immediate action to deal with it'. While emergencies are undoubtedly sudden, their seriousness usually stems from longstanding structural issues and inequities. The emphasis on the suddenness of emergencies often conceals human responsibility. Even disasters, which are typically framed as sudden and as 'natural', cannot be understood without examining the broader context, for example climate change. The destruction following disasters is usually the result of inequalities and poor infrastructure, rather than the actual event itself. Hurricane Katrina illustrates this as the recovery was more deleterious and lethal than the original storm (Adams, 2013; Klein, 2007). The temporality of conflicts and famines is even more long term. The preoccupation with emergency, with the immediate 'present', is one of the reasons why humanitarian interventions do not work. The temporality of humanitarianism prevents organizations from addressing the causes of emergencies (e.g., the causes of famine) and, as a result, humanitarian interventions end up perpetuating the problem (de Waal, 1997). The reason relief organizations avoid looking beyond the temporality of emergencies is because that would implicate them with politics.

The apolitical character of humanitarianism has been promoted as a sacred principle on which the legitimacy of the whole enterprise is premised. This is because only a neutral, impartial and independent organization can be allowed to offer assistance within a nation-state's sovereignty. The apolitical character of humanitarianism is enshrined in the ICRC's principles put forward by Jean Pictet and in particular the principles of 'humanity, impartiality, neutrality and independence', which have influenced the humanitarian field as a whole providing an ethical and operational framework.[2] Humanity means that assistance is directed to all people, regardless of their nationality, gender, ethnicity, race, religion or politics. Impartiality emphasizes that aid must solely be provided on the basis of need, with no discrimination. Neutrality and independence refer to the conduct of humanitarian organizations. Neutrality demands that humanitarian organizations do not take sides in hostilities and do not take action that either benefits or disadvantages any parties in a conflict. Independence means that aid should not privilege or be directly linked to any social group or party to a conflict. Independence has funding implications as the financing of relief efforts should not come from partisan donors (Barnett and Weiss, 2008: 3).

Most humanitarians consider these principles as sacrosanct and an essential part of their identities (Weiss, 2013: 12). However, in practice, the principles are far from straightforward. For example, the neutrality of humanitarian organizations may be key for granting them access to people in need during conflicts, but it can be deeply troubling when it assumes equity between perpetrators and their victims (Terry, 2002: 21–2). This is not a hypothetical scenario. It was for this reason that MSF was established after splintering from the ICRC over its 'neutrality' in the Biafran famine (1967–70). Neutrality has contributed to tragic events such as the killings of Tutsi refugees in the Rwanda camps which also harboured their killers. The presence of humanitarians in the camps created the illusion of safety and ultimately facilitated the genocide that ensued (de Waal, 1997; Terry, 2002). As Rony Brauman, the former president of MSF, has put it: 'neutrality ratifies the law of the strongest' (2000: 107). Neutrality can be harmful. Not taking sides in situations of extreme violence *is* a political decision. Ignoring the causes of famine – which can be interpreted as a neutral stance – can perpetuate the crisis. As Alex de Waal has compellingly argued, the causes of famines are always political. It is an illusion to assume that neutrality is apolitical. Such concerns have led Fiona Terry to refer to the 'paradox of humanitarianism' – the fact that 'instead of alleviating the suffering of the victims, humanitarianism can strengthen the power of perpetrators' (2002: 2).

The principle of independence is equally contested. With some exceptions, most NGOs depend on government funding, which is often determined by political priorities and geopolitical agendas (Rieff, 2002). This explains why some conflicts and emergencies attract a lot of attention and funding, while others remain neglected. In 2024, the tragic conflicts in Gaza and Ukraine dominate the headlines. But there are dozens more conflicts across the world where millions of people are displaced and face extreme hunger and violence. For example, in Sudan 25 million people – over half of the country's population – are in need of humanitarian assistance, with more than seven million people internally displaced since civil war broke out in April 2023.[3] In early 2024, one in three people in Sudan faced acute food insecurity with one child dying every two hours in the Zamzam refugee camp in Darfur.[4] Despite such dire conditions, according to UNOCHA, only 3.5 per cent of the Sudan funding needs were covered in 2024, leaving a US $2.6 billion funding gap.[5] For comparison, the 2024 funding gap in the Occupied Palestinian Territories, including Gaza, was 48.1 per cent according to UNOCHA.[6] Disparities in funding have historic precedents. In 1999, donors allocated US $207 per person for Kosovo, but only US $17 per person in Sierra Leone where, according to objective criteria, the needs of assistance were greater (Terry, 2002: 23). Such inequities directly question whether the principle of independence is actually upheld. These examples reveal the dependency of humanitarian organizations on states, which are more deeply involved in humanitarian affairs than is often acknowledged. We will further explore the relationship between states and humanitarian organizations in the following section.

Empire and the suffering of others

The emergence of humanitarianism is tied to the colonial expansion of the nineteenth century. Imperial expansion through violence caused significant humanitarian suffering. Conflict, famines and epidemics resulted from colonial rule and its policies. At the same time Europeans became aware of the suffering of distant others through colonial institutions such as Christian missionaries (Barnett, 2011). Compassion, advocacy and the imperative to act to alleviate suffering resulted from colonial encounters. In other words, colonialism was both the cause of humanitarian suffering, as well as the vehicle through which this suffering became known to Western publics. A cynical interpretation is that humanitarianism was a way of dealing with empire's guilty consciousness.

There are further ways in which humanitarianism and empire are connected. Humanitarianism served as a justification for colonial expansion. The most extreme example of this was King Leopold who branded himself and his actions in the Congo at the turn of the previous century as humanitarian, thus concealing one of the most brutal and violent colonial regimes in human history (Hochschild, 2019). The case of the Congo captures both how colonialism triggered untold suffering and death; but also, how early humanitarians – missionaries like George Williams – were the ones who exposed the regime's crimes (Hochschild, 2019). Such historical legacies contextualize how contemporary humanitarianism has been used as a veneer for foreign policy, which is why some authors have argued that humanitarianism is used to legitimize an international system that privileges the North and perpetuates global relations of domination (Chimni, 2000; Duffield and Hewitt, 2009; Rieff, 2002; Skinner and Lester, 2012). This happens through the justification of the use of force (e.g., the 'humanitarian intervention' in Kosovo and elsewhere), and the support of neoliberal economic policies spearheaded by transnational institutions such as the World Bank and the IMF which are increasingly involved in development projects and post-disaster recovery. Mark Mazower's history of the origins of the United Nations has revealed that the UN did not simply emerge 'out of ashes of World War II', but was a direct descendant of the legacies of Empire (2014). Even though the UN was later reshaped by the movement of postcolonial independence, some of its founders, such as Jan Smuts, originally saw the organization as a means to protect the old imperial racial order (Mazower, 2014).

Humanitarianism and colonialism are bound together historically, and this relationship explains contemporary iterations of humanitarian practice. The most common way that humanitarianism perpetuates the legacies of Empire is through the persistent asymmetry between aid givers and receivers. Aid continues to depend on the unequal relationship between humanitarian officers, who are typically white, male, and from the minority world, and people affected by crisis who reside in the former colonies of European imperial powers. Terms like 'beneficiaries' and 'donors' invoke the asymmetrical relations involved (Slim, 2015).

Humanitarian interventions typically follow the North–South trajectory, and while there may be occasional South–South interventions there has never been a South–North one. Humanitarianism is a powerful vector for the Western ideas, behaviours and bureaucracies (Donini, 2008: 34). Despite efforts to reform humanitarianism and empower affected communities to hold aid organizations to account, paternalism remains a common criticism of humanitarianism's imperial and contemporary iterations (Barnett, 2011).

Humanitarianism has become a powerful way through which to understand North–South relationships (Donini, 2008). While decolonization entails a radical rethinking of power relations, humanitarianism emphasizes a technocratic approach to recovery. For Donini, humanitarian handouts 'hide the structural issues and force the conceptualization of complex issues into simplistic boxes' (2008: 36). Ultimately, aid is a system that reproduces power inequities between the wealthy North and the poor South – and this system ultimately benefits the North (Escobar, 1995). This is the point where critiques of humanitarianism converge with critical approaches to international development. One difference between humanitarianism and international development is that the former justifies interventions in the name of emergencies.

Both humanitarianism and international development channel aid through organizations based in the North, which benefit from the lucrative contracts. Of course, the structures of power that benefit the North are not just economic, but also political. Aid is often referred to as a form of 'soft power' and is linked to governments' foreign policy agendas. This was evidenced by the decision by the UK government in June 2020 to close the Department for International Development (DfID) and merge it with the Foreign Office.[7] There are countless examples of how humanitarian assistance became entangled with foreign policy objectives. David Rieff discusses how organizations such as the International Rescue Committee (IRC), CARE and others were associated with USA foreign policy during the Cold War (Rieff, 2002: 79–80). We return to our earlier observations about government involvement in humanitarian operations.

Humanitarianism has been historically linked with empires and colonialism. Empires may have collapsed, but the structures through which aid is decided and distributed reveal striking continuities with earlier iterations of humanitarianism. The colonial genealogies of humanitarian relief are reworked in the contemporary context to reveal the tenacity of the geographic, racial, political, social and economic asymmetries between the majority and minority worlds. Capitalism is an additional factor which explains the persistence of these inequities.

Humanitarianism and capitalism

To understand contemporary developments, and in particular the increasing privatization of the humanitarian space, we need to look at the relationship between humanitarianism and capitalism which is not as new as some might assume. It would be mistaken to think that the convergence of humanitarianism and capitalism is a twenty-first-century phenomenon. The relationship between humanitarianism and capitalism goes back to the latter's inception. Capitalism was inextricably linked with colonialism as it was the extraction of resources from the colonies and slave trade that fuelled industrialization and capitalist production. As we have already established, humanitarianism also emerged through colonial expansion and colonial encounters. It was the violence of colonialism that caused suffering, but also colonial encounters that triggered long-distance advocacy and compassion (Barnett, 2011). Capitalism, humanitarianism and colonialism are mutually constitutive of the project of European modernity. Arguably, humanitarianism provided a fig leaf both for colonial expansion and capitalist extraction. The most extreme example of this was King Leopold's brutal colonization and exploitation of the Congo (Hochschild, 2019).

Capitalism, humanitarianism and religion converged into a unique ideology that became part of the 'civilizing mission' of colonialism. All imperial powers justified their expansion and annexation of lands through the notion of a 'civilizing mission' that assumed the supremacy and teleological character of European modernity (Mignolo, 2011). According to Barnett, 'missionaries believed that colonialism and capitalism would civilize local populations' (2011: 66). We can discern a historical assemblage between colonialism, capitalism, humanitarianism and science and technology which has legacies that survive until today.

Traces of these legacies are found in contemporary development projects, and especially those funded by international institutions such as the World Bank, which are increasingly called upon to shape policies in post-disaster or post-conflict areas. The neoliberal policies instituted by such international organizations instil values that may be incompatible with local cultures while they entrench inequities globally. For example, following the so-called 'humanitarian' intervention in Kosovo, one of the first steps was to help the country transition into a market economy through the privatization of state industries (Chimni, 2000). There are additional ways in which we observe the relationship between capitalism and humanitarianism. Disaster capitalism (Klein, 2007) which ensues following emergencies helps to incorporate previously inaccessible areas into the capitalist fold. As Michael Barnett remarked, 'global capitalism needs humanitarianism' (Barnett, 2011: 24).

By addressing social grievances without upending the capitalist order, humanitarianism allows capitalism to expand with a clear conscience. For Marx and Engels 'philanthropists, humanitarians, [and] organizers of charity' [...] ultimately help 'secure the continued existence of bourgeois society' (1968: 58). In the contemporary context, humanitarianism gives global capitalism a fig leaf which explains the growth of 'philanthrocapitalism' and the dynamic entry of the private sector in the humanitarian space, either as donors, partners, vendors or even humanitarians in their own right (Burns, 2019; Barnett, 2022; Henriksen and Richey, 2022). I will further expand on these developments later in the chapter under the 'logic of capitalism'.

The transformations of humanitarianism and the turn to the digital

This brief historical account allows us to identify continuities with the contemporary iterations of humanitarianism. There are also significant transformations such as the increasing bureaucratization and professionalization of humanitarianism. These are the result of the sector's phenomenal growth. In 2023, one in 23 people globally required humanitarian assistance, more than double from 2019. According to an estimate, 570,000 people worked in the humanitarian sector in 2017, which more than doubles the 210,000 figure for 2010.[8] The World Food Programme, the UN's largest agency, increased its expenditure from US $1.2 billion in 1997 to 11.6 billion in 2022.[9]

The sheer size of the field demands significant logistical and administrative capabilities. Humanitarian organizations developed complex bureaucracies and became professionalized as a result of the sector's growth and in response to the genuinely complex emergencies they face. But equally, professionalization and bureaucratization were the result of the sector's need to appear objective, neutral, impartial and apolitical. Bureaucracy standardizes and establishes objective and replicable processes in order to treat all citizens the same way. It is revealing that when the sector faced significant criticisms in the aftermath of the Rwanda genocide, the response was to further bureaucratize and standardize its operations.

The humanitarian sector has grown immensely, but so have the crises facing the world. The funding gap between actual budgets and what is needed is significant. The WFP received a record number of contributions in 2022 (US $14.1 billion) but this fell short of the total amount required to meet all relief needs, which is estimated at US $21.4 billion.[10] The post-cold war world is very unstable with proliferating crises and frequent climate emergencies. At the same time humanitarian organizations strive to appear apolitical and neutral in order to retain their moral authority and legitimacy to intervene. Appearing apolitical is also important from the donors' point of view, so maintaining a 'pure' image of neutrality has funding implications. Aid agencies are also under significant pressure to reform and be accountable to people affected by emergencies. It is in this complex moment of global instability and mounting emergencies, sector growth, combined with immense funding pressures and internal demands for objectivity, neutrality, accountability and transparency that we need to make sense of the growth of 'digital humanitarianism'.

The logic of humanitarian accountability

One of the reasons why digital innovation has been so enthusiastically embraced by humanitarian agencies is because it is seen as a remedy for the deficiencies of humanitarianism. Humanitarianism has been critiqued for reproducing the unequal power relationships on which it is based, for disrupting local societies and creating new dependencies (de Waal, 1997; Terry, 2002). The sector's response to these criticisms has been to launch a series of reform initiatives aiming to standardize humanitarian practice and make it more accountable (Barnett, 2011; Krause, 2014). The interactive character of digital technologies is deemed to address the long-standing demand for humanitarian accountability and reform. Digital technologies are assumed to enable marginalized communities to express their priorities and participate in their relief efforts, thus empowering them to hold aid agencies into account. Similarly, developments in big data and AI have heightened the optimism about the 'participation revolution' and a 'new era of humanitarianism in a networked age' (Meier, 2015; UNOCHA, 2013). Mobile and social media data are assumed to reveal the needs of crisis-affected people thus contributing to a 'rapid decentralization of power' in relief operations (World Disasters Report, 2013).

To understand the turn to digital technologies as a means of improving accountability, it is useful to begin with some key milestones in the history of humanitarian reform from the 1990s. The Sphere Project (1997) and the Humanitarian Accountability Partnership (2003) were established in part as response to the presumed failure of the sector to mitigate the Rwanda genocide in 1994.[11] The Sphere Project was launched by a group of international non-governmental organizations (INGOs) and the Red Cross and Red Crescent movement (ICRC) to improve humanitarian response and strengthen accountability. Sphere established the minimum standards for humanitarian work and enshrined humanitarian principles in its Charter (Krause, 2014: 131). The Sphere Project, or Sphere as it is widely known, was driven by two principles: that people affected by disaster or conflict have the right to life with dignity and the right to assistance; and all possible steps should be taken to alleviate human suffering arising out of disaster or conflict (Sphere, 2018). These principles are enshrined in Sphere's Humanitarian Charter. The Sphere Handbook, which is still widely used and frequently updated – the latest edition came out in 2018 – identifies the minimum standards expected of relief operations in the areas of water supply and sanitation, food security, shelter and health. These standards are measured against specific indicators. For example, the standard of 'hygiene promotion' is satisfied by the provision of 250 grams of bathing soap per person per month (Sphere, 2018).

The Humanitarian Accountability Partnership (HAP) was a multi-agency initiative established in 2003 and set out to establish the sector's accountability principles. This was the first systematic effort to emphasize the importance of accountability to affected people. Of all the ways in which accountability could be defined, HAP chose to equate accountability with feedback, drawing on models from the business sector (Krause, 2014). HAP was replaced by the Core Humanitarian Standard on Quality and Accountability (CHS) which again prioritizes feedback and complaints alongside the provision of information and the encouragement of participation among people affected by crisis.

The 2016 World Humanitarian Summit adopted the 'Grand Bargain', an agreement between some of the largest donors and humanitarian organizations about 'improving the effectiveness and efficiency of humanitarian action' in order to address the widening gap between humanitarian needs and available resources.[12] The 'Grand Bargain' encompassed a series of commitments, including a commitment to a 'participation revolution' and localization, which refers to the increased involvement of local teams into the relief efforts. According to the 'Grand Bargain', the 'participation revolution' is defined by the following: the inclusion of people affected by crises in decision-making processes; the provision of information and facilitation of feedback that is responded to; and ensuring that the voices of the most marginalized people are heard and acted upon.[13]

The legacies of these initiatives are evident in the digital practices that concern us in the book. From being a niche activity in the early 2000s, 'accountability to affected people' (AAP) initiatives are now typically included in all humanitarian projects. Following the 'Grand Bargain' in 2016, AAP and participation initiatives are mandated by the donors. The challenge for aid organizations is how to translate inherently messy processes such as participation into measurable actions. As we'll observe in Chapter 3, accountability to affected people is still largely operationalized as 'feedback'. Interactive technologies have been seized as opportunities to increase the participation of people affected by crisis. This is why several AAP schemes often include a digital component, either in the form of an SMS hotline, a feedback and complaint platform using messaging apps, or even chatbots. Social media and bespoke apps are also used for information. Because of AAP's focus on empowering local communities, even technologies such as biometrics and blockchain, through their uses in digital identity programmes, are often championed as part of the efforts to improve transparency and accountability.

The narrow definition of accountability to affected people as feedback echoes parallel debates in the field of international development. A radical understanding of participation involves recognizing the value of local knowledge and involving people in defining the problems, making decisions and implementing initiatives (Manyozo, 2012). Yet international development agencies were criticized for turning participation into a set of technocratic, measurable goals or even a simple rhetoric in the service of neoliberalism (Cooke and Khothari, 2001). According to Waisbord (2008) the bureaucracy of the sector presented the greatest obstacle for the realization of participatory development. The prevalence of top-down models which allow for greater control of outcomes was incompatible with the radical spirit of participation.

The logic of audit

If the logic of accountability responds to the shortcomings of humanitarianism, the logic of audit stems from its massive growth and expansion. Funding for humanitarian assistance nearly doubled in the space of ten years: from US $16.4 billion in 2012 it grew to US $31.3 billion in 2021 (ALNAP, 2022).[14] To make sense of the growth of the sector it is important to know that states have gradually withdrawn from aid service provision, which they outsource to agencies (Stein, 2008). Governments remain very involved as donors and, in that capacity, demand robust audit trails and evidence of impact in order to justify the cost to their taxpayers. The metrics, data and metadata engendered by digital technologies make them very attractive from an audit point of view.

The demand for metrics needs to be understood in the wider context of the marketization of the humanitarian field. Marketization does not mean that humanitarian organizations operate for profit, but that they conform to market forces in order to 'stay in the business' (Weiss, 2013: 8). Aid organizations constantly compete for funding from a relatively small group of donors (Krause, 2014). The short funding cycle exacerbates the need to have readily available data and metrics to support applications for funding, or the renewal of funding. During my fieldwork, several of my interlocutors told me that they had to apply for funding every few months or even weeks, in order to remain in the field or for their project to remain active. It is worth noting that apart from the costs of the actual relief efforts, humanitarian organizations are now highly professionalized with staff on salaries and large overhead costs. In their funding applications, agencies must provide the evidence of impact that the funders demand. Data, metrics and analytics become this evidence, or the means through which humanitarian organizations continue their relief work, receive funding, and continue to pay their staff. In other words, people affected by crises – through their data – justify and legitimate the aid projects.

Linked to the above is the parallel demand for transparency and efficiencies which is again driven by the donors. The aim of the 2016 'Grand Bargain' was to address the widening gap between ever growing humanitarian needs and the scarce available resources. Improving the effectiveness and efficiency of humanitarianism was the overt goal of the agreement and that priority has been prevalent across the sector since then, if not before. Digital innovation is assumed to be the key way through which efficiencies and transparency can be achieved. Biometrics is a case in point as the technology was introduced in refugee registrations in 2002 in order to address low level fraud, essentially to prevent people from claiming aid twice (UNHCR, 2002). Blockchain technology has been introduced to manage cash transfers in order to cut bank fees. Chatbots have been launched to handle information dissemination and 'communication with communities' (CwC) in order to free up staff time when resources are stretched.

The above observations are not unique to the field of humanitarianism. There has been an explosion of audit in all spheres of public life, from schools and universities to hospitals and government departments (Strathern, 2000). Similarly, the push for efficiencies is also widespread across all sectors. Neoliberal policies which became popular in the 1980s have affected humanitarianism just like they have shaped other aspects of social life. It is, however, erroneous to attribute all these transformations to neoliberalism. As Monika Krause (2014) has observed, the push for the efficient management of humanitarian organizations emerged well before the 1980s. Because relief operations involve complex logistics, the aid sector developed methods to address that. One of these methods was the logframe, which is essentially a table, or matrix that traces the development of projects and measures their success against predetermined, measurable indicators. We saw what these indicators look like when we examined the Sphere project in the earlier section on the logic of accountability. The example I offered there was the indicator for the standard of hygiene promotion: 250 grams of soap per person per month indicates that the standard has been achieved. By setting objectives, action points, indicators and deliverables, the logframe is an example of bureaucratic rationality that not only manages relief work, but also shapes it. If the neoliberal turn in the 1980 simply accentuated features of a system that was already in place, digital technologies and computation are now taking it to the next level.

The latest development in the professionalization of audit is the introduction of third-party monitoring (TPM) where a private company is contracted to assess the performance of a humanitarian project, usually at the donor's request. This is a growing industry in the aid sector with over $200 million spend on TPM in Afghanistan between 2006 and 2016 (Diepeveen et al., 2022). Humanitarian data collection and management reveals a complex supply chain of private contractors, donors, humanitarian agencies and data subjects themselves (Diepeveen et al., 2022). TPM practices raise significant questions about data safeguards, especially as the private companies lack knowledge of the sensitive context. I'll return to this theme in the next section. TPM also reveals how the logic of audit intersects with the logic of capitalism.

The logic of capitalism

In February 2019 the United Nations World Food Programme and Palantir Technologies signed a partnership which caused a great controversy given Palantir's track record in intelligence and military projects through its association with ICE and the CIA. There were many concerns expressed about the privacy and data safeguarding of some of the world's most marginalized people. Would Palantir have access to the data or metadata of the over 91 million people who receive assistance from the WFP each year?

The WFP–Palantir relationship may have attracted a lot of attention, but it is just one example of what has become the 'new orthodoxy' in the humanitarian space (Sandvik et al., 2017). There are thousands of private–public partnerships in the aid sector, many of which involve technology companies. Accenture, Amazon, Google, Meta, Microsoft are only some of the companies that collaborate on aid projects with humanitarian organizations. Cindy McCain, the executive director of the WFP, in a 2023 speech aiming to attract further private-sector investment declared that: 'the humanitarian sector is one of the world's biggest growth industries'.[15] Some companies, including start-ups, often launch their own small-scale humanitarian projects in the shape of a chatbot or an app for refugees. Then there is the phenomenon of 'philanthrocapitalism' where billionaires set up foundations to address the world's problems (Burns, 2019; Bishop and Green, 2008). The private sector is also involved as vendors to whom humanitarian organizations increasingly outsource projects, such as biometric enrolments or cash transfers. Such an example is IrisGuard, a biometric technology company involved in border securitization and humanitarian cash distributions.[16] We already observed the practice of third-party monitoring, which is the outsourcing of the evaluation of humanitarian projects to private companies.

For commercial companies, the involvement in humanitarian projects represents excellent branding and public relations opportunities. Humanitarian projects can increase the visibility of companies, offer them access to new markets and lucrative data, and the opportunity to test new technologies. For technology companies like Meta, the parent company of Facebook, which has received widespread criticism about its business model and involvement in the Cambridge Analytica scandal, humanitarian projects can compensate for such negative publicity.

A well-known critique of corporate social responsibility (CSR) is that by shrouding the market in a moral discourse, it provides a veil for profit-making. However, my fieldwork reveals that the motive for profit is openly acknowledged in industry events and the popular press with headlines such as 'Europe's Refugee Crisis is a major opportunity for businesses', or 'Refugee camps are an untapped opportunity for the private sector'.[17]

It is important to note that the logic of capitalism does not just concern companies based in the global North. Companies in the global South have internalized the logic of capitalism and express it openly. This became very evident during my fieldwork of the aftermath of Typhoon

Haiyan in the Philippines. Two of my interlocutors from a local technology company were upfront about how the company saw a 'clear business opportunity' in Haiyan. Within days of the Typhoon making landfall, they arrived in Tacloban, the city most severely hit by the Typhoon, and started a humanitarian connectivity programme disseminating 'free' SIM cards to people who had lost their phones during the storm. As a result, the company was able to increase its market share in the region by a significant margin. The hashtag from the project accompanied the company's marketing campaigns for over one year after the Typhoon. Here, disaster capitalism (Klein, 2007) is not veiled, but in fact expressed unequivocally.

The phenomenon of 'philanthrocapitalism' deserves some attention as it differs from CSR and other traditional forms of philanthropy. Philanthrocapitalists such as Bill Gates do not just provide funding, but decide on the problems, the approach and the solutions. Bill and Melissa Gates have set up the eponymous Foundation which is so large that it accounts for 'more than half of all global philanthropic giving to development today' (Fejerskov, 2022: 90). The Bill and Melissa Gates Foundation has invested over US $71 billion in grant payments from its inception to 2022 and has set the agenda in the areas of global health, education, financial services for the poor and gender equality among others.[18] The beginning of philanthropcapitalism can be traced to Ted Turner, who in 1997 made a massive donation of US $1 billion to the UN to set up the UN Foundation, a strategic partner to the UN which builds public–private partnerships to support the work of the UN (Bishop and Green, 2008).

Behind some of these initiatives is the assumption that the private sector is agile compared to the cumbersome bureaucracy of organizations such as the UN and is therefore able to innovate. This is a common refrain going back to the 'Californian Ideology' which held that entrepreneurs know how to innovate and should therefore be allowed to fix the world's problems (Barbrook and Cameron, 1996). By contrast, several of my interviews with the private sector alluded to the 'sclerotic UN system' and its burdensome bureaucracy.

The privatization of humanitarianism raises critical questions. There is concern about whether private companies abide by the principles of humanitarianism and the imperative 'do no harm'. Unlike UN agencies and NGOs, private companies are under no obligation to do so. Further, there are no clear pathways for accountability or redress if something goes wrong. Another important aspect of this development is how private corporations use the opportunities for humanitarian work to extend their authority over the social order (see also Rajak for a parallel discussion on CSR, 2011). Through their corporate social responsibility and foundation departments and through private–public partnerships, businesses pursue their own agendas. By turning themselves into agencies for humanitarianism, corporations reframe political problems in line with their commercial interests. To render displacement, which is a political problem, an issue that can be 'solved' with 'mobile connectivity', is to depoliticize displacement while advancing business agendas. The logic of solutionism further contributes to the depoliticization of humanitarian crises.

The logic of technological solutionism

Technological solutionism, or technosolutionism, refers to the desire to find technological solutions for complex social problems. It is closely linked to the logic of capitalism as its greatest advocates come from the private sector. In fact, the roots of the phenomenon are found in the Californian Ideology and the belief that entrepreneurs and digital developers are the ones who have the solutions (Barbrook and Cameron, 1996). Solutionism epitomizes the neoliberal principle of the small state and market ingenuity. There is an unshakable belief that there are uncomplicated solutions for even the most complex problems and these solutions are almost always technological in nature. The apotheosis of this mantra can be found in the optimistic narratives around AI as a potential solution to everything: from climate change to world hunger.

Technological solutionism equally originates from the legacies of the countercultural movements of the 1960s, which shaped ideas of digital utopianism in the 1990s 'cyberculture' (Turner, 2006). In this romantic imaginary of technology, the 'hacker' is cast as a 'romantic hero', a charismatic figure who codes for a better future (Streeter, 2011). The imaginary of technology as a form of enchantment (Gell, 1992; Madianou, 2021), spirituality (Supp-Montgomerie, 2021) and charisma (Ames, 2019; Mazzarella, 2010) has a long history. As Langdon Winner observed 'since the early days of the industrial revolution, people have looked at the latest technology to bring individual and collective redemption' (1997: 1000). This is a teleological, but also an ideological imaginary of technology the latest iteration of which is found in the optimistic discourses about AI.

The most critical aspect of technological solutionism is that it puts solutions before problems. As one of my interlocutors from the aid sector put it, 'rather than seeking solutions for problems, solutionism seeks problems for solutions'. The concern here is the foregrounding of solutions, before the understanding of the actual problems or cultural contexts. This is illustrated in the following quote by another interlocutor from the sector:

> Now two years ago [...] nobody in the humanitarian sector was talking about blockchain. [...] Now you go to meetings and you get people saying we want to try something with blockchain and then you probe it a little bit and they don't really understand what blockchain is. And they don't really understand what value they want to get out of it is. They just know it's an innovation and they've heard there's something so they want to give it a try. So it becomes, you know, that's a sense in which a specific technology which is perceived as innovative becomes cover perhaps for things which don't necessarily need that technology to be done. So there are lots of ways in which you can use a blockchain but blockchain is not the only way of doing those things. There are other distributed databases. There are other modes of encryption. And you don't have to use blockchain to get the benefits of those. But because blockchain has a high profile because it's at the peak of the Gartner Hype Cycle, that's what people are focusing on.

Starting with a technology – in the above example, with blockchain – and then finding a problem to apply it to is problematic in all contexts, let alone in situations of great precarity and vulnerability such as those experienced by crisis-affected people. Starting with a solution, means that understanding the issues people face becomes secondary. The goal of technological solutionism is not to make sense of the actual problems and their socio-political context, but rather to code and to test technologies. Lilly Irani observes that hackathons are informed by the 'pedagogies of computer science', which privilege a 'bias to action' (2019: 125). Hackathons almost never question the parameters of the project or the power relations at stake. The objective is to produce a functioning code, which reflects the training of engineers and computer scientists, which resembles a 'long chain of hackathons' (Irani, 2019: 125).

It is not just entrepreneurs and hackers who favour technological solutionism. The logic resonates with the practices of government and international institutions involved in international development such as the World Bank. Tania Li observes that international development projects operate by identifying problems and then 'rendering them technical' (2007: 7). To render a problem technical is to make it intelligible and therefore solvable. By reducing problems to technical fixes, international development projects depoliticize aid and the power relations therein. It is for this reason that James Ferguson referred to international development as the 'anti-politics' machine (1994). The argument is entirely applicable to humanitarian projects. While neither Ferguson (1994), nor Li (2007) addressed digital technologies, their arguments acquire renewed urgency in the contemporary context as existing practices in international development and humanitarianism converge with the solutionism of the technology industry.

We can discern here parallels with other sectors such as education (Sims, 2017) and events such as the response to the Covid-19 global pandemic (Madianou, 2020; Milan, 2020). One major difference between technosolutionism in humanitarianism and international development compared to other sectors is that the former involve power inequities between the global South and North. These global inequities, which map onto colonial genealogies, are further entrenched by the application of the solutionist paradigm.

The logic of solutionism explains the prevalence of hype and technological experimentation in digital humanitarianism projects. This is evident in the discourses about 'disruption' and 'failing fast' which I encountered in industry events and hackathons (Madianou, 2019a). The reference to 'disruption', echoing in the Silicon Valley mantra 'move fast and break things', is particularly troubling when applied in some of the most fragile settings in the world. The concern here is that such a cavalier approach is incompatible with the 'do no harm' imperative of humanitarianism. Yet we see that refugee camps and other crisis settings are treated as laboratories for experimentation with untested technologies (Jacobsen, 2015; Madianou, 2019a and b; Molnar, 2020) – a theme that we'll further explore in Chapter 4. Experimentation reveals, but crucially also entrenches the power geometries of humanitarianism and the 'technology for good' fields.

The decoupling of understanding and action within the logic of technological solutionism exemplifies Lilie Chouliaraki's notion of post-humanitarianism, which refers to solidarity without the moral and emotional weight that accompanies the deep engagement with distant suffering (Chouliaraki, 2013). The self-referential dimension of post-humanitarian action becomes evident in recent critiques of hackathons as the spaces for the production of entrepreneurial subjects (Irani, 2015) and as 'a great way to build networks, strengthen communities and reinforce beliefs in common goals' rather than actually help displaced people (Geber, 2016). Hackathons exemplify the post-humanitarian disposition of 'feeling good' about oneself. Coupled with the logic of capitalism, solutionism leads to the normalization of technological pilots or experiments among people made vulnerable by crises or displacement.

The logic of securitization

At first glance, the logic of securitization primarily involves the role of the state. As we have already observed national governments are directly involved in relief efforts as donors, or as hosts to refugees. Governments are also involved in relief operations when emergencies occur in their territories. Humanitarian agencies operate at the invitation of national governments and therefore are required to cooperate with them. The logic of securitization applies primarily in the response to migrant and refugee flows. Migration post 9/11 has been framed as a security problem (Bigo, 2002) as exemplified in the refugee crisis of 2015 when European countries responded by closing their borders. The logic of securitization reduces refugees to a security threat (Andersson, 2014: 68), not because migrants are such a threat, but because migration has been ideologically framed in that way.

The securitization of migration is highly technologized – in fact the levels of securitization currently in place would have been impossible without the massive expansion of the security–industrial complex since 9/11. The phenomenal growth of the biometric sector since 2001 is a case in point. Biometric technologies are the most common method through which governments make populations legible (Scott, 1998) in order to detect anomalies and control borders (Aradau and Blanke, 2017). Biometrics 'inscribe the boundaries between safe / dangerous, civil/ uncivil, legitimate travel / illegal migrant' (Amoore, 2006: 337). Increasingly, a whole assemblage of technologies is employed for border surveillance, from drones to AI (Molnar, 2020). In parallel, we are witnessing a 'border externalization' which refers to the extension of the border and migration controls into neighbouring countries or sending states in the global South (Napolitano and EuroMed Rights, 2023; Walia, 2021). Externalization works when powerful states (such as the EU or the US) request that sending or transit countries pre-emptively control the flow of migration through bureaucracy (e.g., issuing of visas), or surveillance and interception (e.g., through the use of drones). The power asymmetries between countries in the rich North and the poorer South give rise to what Harsha Walia calls 'border imperialism' (Walia, 2021). By border externalization European countries also outsource responsibility and blame host countries when atrocities happen (Walia, 2021: 100). The establishment of common electronic databases and infrastructures of information sharing is one of the ways through which border externalization takes place.

The other side of the coin involves the internalization of borders. The policing of migration is increasingly outsourced to institutions such as the health service, schools and universities among others who are required to check migrants' legal documents under the threat of deportability. Digital technologies and datafication multiply the geographical border and the opportunities for remote control (Dijstelbloem and Broeders, 2015). Biometric technologies and interoperable databases play a key role in the internalization of borders. Eurodac, the European Dactyloscopy Database, is key for enforcing the Dublin Regulation which determines the EU country where a refugee can process their asylum claims (usually the country of entry in the EU). If the refugee's fingerprints are recorded in a different country they are deported to the original country of entry (Metcalfe and Dencik, 2019). As Louise Amoore remarks, with biometric technologies the migrant 'body becomes the carrier of the border' (Amoore, 2006: 336–7). The permanent records afforded by biometric technologies enable the surveillance of refugees in perpetuity.

The fact that border technologies involve vast, permanent and interoperable datafied systems have led authors to theorize them as infrastructures (Dijstelbloem, 2021). Although the logic of securitization concerns the border *par excellence*, it also applies to settings such as refugee camps where the use of biometrics is completely normalized. By infrastructures of securitization and control I include the technological systems of the border, as well as the systems of surveillance and containment in refugee camps and other humanitarian operations.

These infrastructures of securitization and bordering are not neutral, but rather produce race and other forms of marginalization. Infrastructures depend on classifications in order to function. Classifications are embedded in the infrastructure, and are therefore hidden, but they have powerful political consequences (Bowker and Star, 1999: 319). Classifications are always contested and reflect social biases. This is exemplified by biometric technologies which are based on classifications regarding the measurement of the human body that have colonial genealogies. For example, biometric technologies assume 'prototypical whiteness' in their design, and are prone to high margins of error when recognizing othered bodies (Benjamin, 2019; Magnet, 2011; Pugliese, 2010). The ways biometric technologies produce race is the focus of Chapter 2. But for the purposes of this discussion, it is important to recognize that 'border regimes variously allocate and curtail mobility and migration on a racial basis' (Achiume, 2021: 333).

Securitization is an area where the synergies between states, humanitarian organizations and private companies become apparent. The use of biometric technologies helps illustrate this. We have already seen that humanitarian organizations use biometrics to address internal priorities, such as pressures to address low-level fraud, establish robust audit trails and make efficiencies. Further pressure for aid agencies to use biometrics comes from the governments of large donor countries like the US (The Engine Room and Oxfam, 2018). The interoperability of biometric systems make them particularly attractive to governments. It is known that governments have data sharing agreements with intergovernmental agencies such as UNHCR (Jacobsen, 2015 and 2022). Given UN agencies and INGOs operate under the jurisdiction of host states, there is little room to refuse a government request. Furthermore, UNHCR often conducts biometric registrations together with the host country as we saw in the Rohingya example in the Introduction. The sharing of sensitive data raises concerns about 'function creep', which refers to the way that data collected for one purpose (e.g., refugee registration) may end up being used for an entirely different purpose (e.g., state surveillance) (Ajana, 2013; Taylor, 2016). All this is complicated by the involvement of private firms in the development and management of biometric technologies and infrastructure. As the agendas of the different actors converge, we observe a heightened demand for data (Madianou, 2019b; Lemberg-Pedersen and Haioty, 2020).

The logic of resistance

As always, structures of control produce contestation. The sixth logic is resistance, which refers to the extent to which people challenge practices of control, datafication and automation. 'There is no power, without potential refusal or revolt', Foucault reminds us (2000: 324). He urges us to move beyond the binary thinking of 'either intransigence, or a blanket acceptance' as these typically co-exist (Foucault, 2000: 455–6). If we turn to the authors of postcolonial theory (Said, 1994) and the Black radical tradition (James, 1938; Patterson, 2022; Robinson, 2021) it is evident that resistance took place even in the most oppressive environments such as slave plantations. In Chapter 6, I develop the notion of 'mundane resistance' to capture the ways in which people's agency and everyday contestation is expressed in asymmetrical settings.

Much of literature on the securitization of migration as well as the literatures of datafication and surveillance do not take into account people's lived experiences. As a result, refugees or people affected by crisis remain a blank category which reinforces popular perceptions which cast them as lacking agency. Lilie Chouliaraki and Myria Georgiou argue that the digital border cannot be solely understood as a site of control and surveillance. It is also a site of conviviality, solidarity and hope for a better future (Chouliaraki and Georgiou, 2022). Social media, mobile phones and biometric systems together with other technologies constitute an infrastructure of movement as well as control (Latonero and Kift, 2018). This infrastructure allows refugees to complete their perilous journeys and even to circumvent surveillance (Gillespie et al., 2018). Yet the same infrastructure is also used for surveillance, control and discrimination.

Scholars working within the autonomy of migration approach offer a more radical analysis of migrants' agency. The securitization of migration is a response to people's agency and not the other way around: it is because people have agency that structures of control are imposed on them (Papadopoulos, Stephenson and Tsianos, 2008: 43). Migrants' capacity to subvert border controls is well documented (Papadopoulos, Stephenson and Tsianos, 2008; Scheel, 2019). Examples include the forgery of legal documents including passports, the physical destruction of border fences or the damaging of fingerprints so as to evade biometric border control (Scheel, 2019). Papadopoulos, Stephenson and Tsianos (2008) document how migrants escape control and create social change by becoming imperceptible to the political system of global North Atlantic societies. While I share the emphasis on people's agency, the question of whether everyday resistance precedes or responds to power and control misses the point that these are relational processes. Even if securitization is a response to migrants' agency, the policing of borders and refugee camps will in turn generate further resistance. Rather than prioritizing one over the other, the book examines resistance and power as mutually dependent.

In order to understand practices of resistance, the book develops an ethnographic approach to infrastructure – as opposed to an approach that only focuses on the analysis of technologies. An ethnographic approach to infrastructure is well suited to reveal the relational and, ultimately, contested meanings of technologies and data in emergency settings.

Digital infrastructures are tools of control, but they also offer opportunities for sociality, personal fulfilment and even resistance. To acknowledge resistance should not detract from the argument regarding power. People also appropriate technologies in unpredictable ways and invest them with meanings.

There is a long-standing trend in the field of media and communications for paradigms to oscillate between those that favour media power to those that favour powerful audiences.[19] We observe similar trends in the area of internet studies. In the 2000s the field of internet studies was characterized by an initial optimism regarding the potential of digital media to revolutionize political participation and almost all spheres or social life. Authors emphasized the power of networked communication for social movements and protest (Castells, 2012 among others). But soon the pendulum swung in favour of powerful technologies. Edward Snowden's revelations about the sweeping surveillance practices by the US and other states combined with the increasing realization that the business model of social media depends on the selling of users' data for profit (Lyon, 2015; Zuboff, 2019). Scholars foregrounded in their analysis the datafication and automation of all aspects of social life, from welfare to policing – and the fact that these algorithmic systems reproduce and amplify racial, class and gendered inequalities (Benjamin, 2019; Dencik et al. 2022; Eubanks, 2017 among others). Yet to acknowledge the power of algorithmic systems, should not allow us to lose sight of the fact that digital platforms are often appropriated in oppositional ways. In fact, to ignore that would amount to technological determinism. Rather than having to choose between media power or people's agency, the book focuses *both* on the power of infrastructures (as well as technologies and practices of computation) *and* people's agency. Both exist in parallel, even though they are unequally distributed, especially in the asymmetrical situations such as those that we encounter in this book.

Conclusion

Humanitarianism is a fragmented field, involving different actors motivated by different logics. The chapter outlined the six competing logics (accountability, audit, capitalism, solutionism, securitization and resistance) that drive the digitization of humanitarian operations. The distinction between these logics is largely analytical. In practice the logics intersect to produce the phenomenon of technocolonialism. Interestingly all actors in the field, from the state and humanitarian organizations to private companies, are interested in digital technologies and data albeit for different reasons. This explains the increasing digitization and datafication of humanitarian operations. Accountability, participation, audit and security have been 'rendered technical' (Li, 2007) with associated technical fixes. As was evident in several of the examples discussed so far, these logics converge. For instance, as audit processes are outsourced to private companies (for third-party monitoring or biometric cash assistance) the logics of audit and capitalism merge. IrisGuard, a biometric technology company which provides the infrastructure for the UN World Food Programme's flagship cash assistance scheme, is also involved in border securitization.[20] Here the logic of securitization combines with the logics of audit, capitalism, solutionism (the biometric systems 'solve' low-level fraud), and even accountability (biometric technologies speed up distributions and make aid more equitable). As biometric technologies are inevitably contested, the logic of resistance also applies. Chapters 2–6 will draw on the analytical framework of the competing logics in order to trace the consequences of the datafication and digitization of humanitarian operations.

The discussion so far has questioned the assumption that humanitarianism is apolitical. From its inception humanitarianism was caught up in politics and this continues today. Politics is present at all levels: the relationships with host governments, the military, relationships with donors (typically the governments of rich nations in the minority world), the priorities of funding bodies, including the private sector, the partnerships with powerful foundations and, of course, the relationships with local populations. Humanitarianism is defined as apolitical and neutral, but it is impossible to disentangle humanitarianism from politics. A note to make here is that 'apolitical', too, entails a political stance – something that humanitarians were confronted with during the Rwanda genocide. In their effort to remain neutral, relief agencies ended up endangering the lives of those whom they were meant to protect (Terry, 2002).

In digital technologies and AI, humanitarian organizations see opportunities to further distance themselves from politics and present themselves as 'pure'. The association of computation with science and, therefore, with objectivity, is why technologies such as biometrics and automated decision-making are very attractive to aid organizations. Such a view ignores the fact that technology is never neutral. Computation depends on classifications, which are always subjective (Bowker and Star, 1999). As we have already argued the classifications of biometric technologies discriminate by design as they impose racialized, gendered, ageist and ableist categories on body measurements. Chapter 2 will unearth the colonial genealogies of biometrics and argue that their normalization produces harms which exemplify technocolonialism.

As digital technologies and computation emerge as inherently political, what are the implications for the humanitarian principles of humanity, impartiality, neutrality and independence and which we discussed in the beginning of the chapter? Can the principles of humanity and impartiality – which refers to the fact that aid should be directed to all humans based on their needs regardless of their race, gender or ethnicity – be upheld, if biometric technologies discriminate by design? What are the implications for the principle of independence when the infrastructures of aid are increasingly privatized and owned by companies which are answerable to shareholders? And how can the principle of neutrality be achieved if technologies are known to systemically marginalize some groups? These are some of the core questions that Chapters 2–5 will explore before we return to them in the concluding chapter.

2

Biometric Infrastructures

Once a month the aisles of Tazweed, a supermarket in the Za'atari refugee camp in North Jordan, are very busy with customers. Camp residents have received their monthly food allowance of 23 Jordanian dinars (US $32). Fatima, a mother of three children, peruses the shelves and compares the prices for eggs, milk and oil. At the checkout counter, instead of paying with cash or credit card, she gazes into a black box with a camera at its centre. The camera scans her iris and, after a while, the merchant receives a message that payment has been received.

Fatima is one of the 80,000 Syrian refugees living in Za'atari.[1] Like most camp residents she receives aid through the World Food Programme's (WFP) *Building Blocks* scheme which uses biometrics and blockchain technology. Once Fatima receives an SMS that her aid entitlement has arrived, she can use it in one of the two designated supermarkets in the camp. When Fatima scans her iris, her details are authenticated against the United Nations High Commissioner for Refugees (UNHCR) database. Once authentification is completed, the WFP blockchain technology releases an electronic payment to the merchant.

Building Blocks started as a technological pilot in Jordan in 2016, but has since been rolled out in Bangladesh, Lebanon, and since 2022, in Ukraine. It has reached over four million people and has processed over US $529 million in cash transfers between 2016 and 2023.[2]

Biometric technologies have become ubiquitous in the humanitarian sector. Biometric technologies are not just used in the mandatory biometric registration of refugees; they are increasingly integral in everyday practices such as cash distributions and digital identity programmes which use technology to identify and verify people. More broadly, biometrics also underpin the infrastructures of migration control and securitization. Recent years have seen not only the normalization of biometrics (Jacobsen, 2015), but also the acceleration of the rate of biometric registrations, which becomes a bureaucratic goal in its own right. Because of its infrastructural and pervasive character, biometrics is the technology that most exemplifies the increasing digitalization and datafication of humanitarianism, which is why we begin our exploration with this particular socio-technical assemblage.

Biometrics is a technology for measuring, analysing, and processing a person's physiological characteristics, such as fingerprints, iris, facial patterns, voice, hand geometry, and DNA among others. Contemporary biometrics rely on computational developments, including AI, but they are far from a new phenomenon. The genealogy of biometrics can be traced to the nineteenth-century practices of bertillonage and the uses of fingerprinting for the control and management of colonial subjects in the British Empire and (Breckenridge, 2014). Unpacking the genealogy of biometrics reveals the inextricable way they have been linked with empire, subjugation and racial segregation. The chapter starts with this history, before exploring how contemporary iterations rework these links. The chapter draws on the interviews and digital ethnography of the 'digital humanitarianism project'.

Biometrics, race and empire

Biometrics has always been a technology of race and empire. Contemporary biometrics stems from the now discredited subjects of anthropometry and phrenology in the nineteenth century (Magnet, 2011; Pugliese, 2010). The main identification system before fingerprinting was invented was bertillonage. Named after Alphonse Bertillon, a French police officer, bertillonage was used to identify criminals based on physical measurements such as the size of someone's skull. Bertillonage drew on the flawed principles of anthropometry and phrenology, which are steeped in racist, classist and gendered assumptions about human bodies. Such problematic assumptions informed all iterations of early biometric methods. Frances Galton, the inventor of fingerprinting and one of the founding figures of statistics, was an outspoken advocate of eugenics. Galton shared the principles of Social Darwinism and in particular the idea that Empire was justified because of the racial supremacy of the colonizers as part of human evolution (Breckenridge, 2014: 38). We see one yet example that uses 'science' to justify violence and the subjugation of people.

Fingerprinting was developed by Francis Galton, but was implemented in nineteenth-century India by the colonial authorities of the British Empire. Galton's observations about the unchanging nature of fingerprints led him to develop an identification system that found its perfect home in colonial India and South Africa. Following the 1858 mutinies in Delhi, the East India Company handed the administration of India to the British government. The British sought ways to assert their authority and address problems like fraud. Sources from the period reveal that officials suspected that Indian people claimed pensions twice by impersonating dead relatives (Cole, 2001: 64). The introduction of fingerprinting was justified on the basis that local people could not write their names while they appeared 'indistinguishable' to the British administrators. In 1891 Galton wrote:

> it would be continual good service in our tropical settlements, where individual members of the swarms of dark and yellow skinned races are mostly unable to sign their names and are otherwise hardly distinguishable by Europeans [...] and whether they can write or not, are grossly addicted to personation and other varieties of fraudulent practice. (Cited in Breckenridge, 2014: 63–4)

By 1890 fingerprinting was adopted across the colonial administration in India – from the payment of government pensions to medical examinations (Breckenridge, 2014: 67). It didn't take long before fingerprinting was introduced in other British colonies such as South Africa where mining corporations sought a system of managing and controlling indentured migrant labourers. Mahatma Gandhi famously protested against the compulsory fingerprinting of Indian labourers in the Transvaal leading to his arrest. This was one of his first clashes with the British Empire. Reacting to the discriminatory practice of fingerprinting, Indian labourers said that it felt like 'having a dog collar put on them' (Singha, 2000: 194). Gradually, the universal ten-fingerprint registration was introduced to the whole of the Transvaal, although, in practice, 'universal' meant Black people. White constituents would only be fingerprinted if they had a brush with the law (Breckenridge, 2014: 86). Fingerprinting became a key tool for entrenching racial segregation.

Simone Browne (2015) has compellingly analysed the practice of branding in the transatlantic slave trade as a form of biometrics. Marking the bodies of slaves was both a brutal method of corporeal punishment and a method of identification which traced slaves to their owners (Browne, 2015). Branding and tattooing were also widely used in the Indian penal colonies which were established by the British (Anderson, 2004; Singha, 2000). Between the late eighteenth and early twentieth centuries colonial authorities sent tens of thousands of Indians to penal colonies across the Indian Ocean Archipelago and South East Asia (Anderson, 2004). Branding the bodies of the enslaved and the imprisoned is the most violent manifestation of the essence of biometrics which uses the body as a method of identification. In these cases, branding simultaneously objectified bodies and subjugated people. The body was indelibly marked as chattel, criminal or deviant.

We observe the deep-rooted link between fingerprinting, imperialism and racism. Fingerprinting was an identification technology only reserved for suspect bodies: slaves, criminals, migrants, or natives. Those deemed to be respectable citizens were not required to be fingerprinted. Indeed, written forms of identification and contracts became dominant in countries like Britain, where fingerprinting was never introduced and where 'identification through the body was associated with the nonrespectable, the deviant and the alien' (Higgs, 2011: 77). The resistance to fingerprinting in Britain during the twentieth century has been traced to the association of fingerprinting with criminality, but also the public outcry regarding the controversial fingerprinting of Indian migrants in South Africa (Higgs, 2010: 64).

By contrast, fingerprinting continued as a practice in India and South Africa even after the collapse of the British Empire. As is often the case, postcolonial states, through the newly formed class of bureaucrats, adopted the policies of empire. The internalization of imperial practices of classification and enumeration by the colonized engendered new racial hierarchies and explains the continued presence of colonialism well after formal end of colonial rule. It is perhaps no coincidence that India currently runs Aadhaar, the largest biometric digital identity system in the world (Singh and Jackson, 2021; Rao and Nair, 2019).

While there have been technological advances in the field of biometric technologies, some patterns remain very similar to that early period. Despite the clear parallels between early and contemporary biometrics, there are also differences. Before exploring these comparisons let's examine the contemporary iterations of biometrics.

The biometric–industrial complex

The contemporary biometric industry grew out of the US prison–industrial complex in the mid- to late twentieth century, which further underscores the link between biometrics, criminality and the disciplining of marginalized populations (Magnet, 2011). The biometric industry exploded after 9/11, which, among other things, was seized as a terrific business opportunity by the prison–industrial complex (Magnet, 2011). 9/11 marked the 'realignment of national security interests with the profit of private companies' (Monahan, 2010: 37). The Covid-19 pandemic between 2020 and 2022 further accelerated the growth of the sector. The social distancing measures mandated by public health policies explain the increased demand for contactless transactions enabled by face recognition technology. The biometrics sector was valued at US $42.9 billion in 2022 and is predicted to reach US $84 billion by 2026.[3]

Contemporary biometrics extend to a wide range of physical (e.g., fingerprints, iris patterns, face geometry, hand geometry, vein patterns and DNA among others) and behavioural (e.g., gait) attributes which are referred to as modalities. Voice is also an increasingly popular biometric modality which is a hybrid physical and behavioural attribute. The range of modalities reflects developments in computing, biological sciences and physics among others. Biometrics combines a range of technological methods: from machine learning and AI, to blockchain, as was evident from the example which opened this chapter. In the next paragraphs I will unpack the features of these systems.

Biometric data are used for identification and verification purposes. Identification checks a record against a large database of biometric profiles (one-to-many comparison), whereas verification checks a live record against the entry already in the system (one-to-one authentication). Identification processes entail a higher risk of false matches than verification (The Engine Room and Oxfam, 2018). Humanitarian agencies commonly use biometric data for identification and authentication purposes.

The capturing and processing of biometric data depend on automated systems of algorithmic sorting. Similarly, biometric identification occurs largely through artificial neural networks (ANN) that employ machine learning algorithms to process complex data entries which learn to imitate the function of the human brain, by recognizing patterns or shapes (Bowyer et al., 2008). Several factors determine the success rate of the method, including the type of algorithms used for processing (e.g., segmenting), or indexing the iris scan, as well as the algorithms and input data used for training the neural networks (Bowyer et al., 2008). Research has shown that the datasets used for the training of face recognition algorithms have a white, male bias and therefore have significantly higher margins of error when encountering darker female subjects (Buolamwini and Gebru, 2018). Underpinning all these technical processes are human decisions regarding classifications (Bowker and Star, 1999).

Recently, as illustrated by the example which opened this chapter, biometrics have been combined with developments in blockchain, which is best known as the technology behind bitcoins but has a wider set of applications. Blockchains are distributed ledgers, or shared databases. Any participant on a blockchain network can submit and review 'blocks' of information in real time. For example, when a participant in the *Building Blocks* scheme records a new transaction, this is automatically replicated on all system nodes following biometric and algorithmic verification. The network constantly reconciles information, so that all users access the most up-to-date version of the blockchain. The distributed nature of information means that even if one node is shut down, the network will not be affected. At the same time, information cannot be deleted; new blocks can only be added. This – often praised – immutability of blockchain can have negative consequences if data entries are erroneous.

Although public blockchains are better known (as they are used in cryptocurrency systems), private blockchains are also common and often preferred by humanitarian organizations as they only allow access to information to those granted permission. The WFP *Building Blocks* scheme was initially launched on a public blockchain, but scalability problems relating to speed and cost shifted the project to a permissioned blockchain (Juskalian, 2018).

As blockchain, AI and biometrics combine into interoperable systems it is important to try to make sense of their convergence – as well as the distinctive characteristics of each constituent technology. Machine learning algorithms amplify existing risks associated with biometric measurements, storage, and identification processes. Big data are used to train ANN algorithms. Blockchain-enabled cash transfers use biometric verification, which depend on algorithms. The replicability and public nature of blockchain ledgers raise questions about the privacy and protection of sensitive data, whereas blockchain's immutability can make an erroneous record permanent, which can have severe consequences for displaced people (Coppi and Fast, 2019; Madianou, 2019b).

Biometrics in the humanitarian sector

The United Nations, and in particular two of its agencies, UNHCR and the WFP, has been driving the expansion of biometric technologies across a range of humanitarian operations. Biometric technologies were first introduced in 2002, when UNHCR piloted iris scans in the registration of over 1.5 million Afghan refugees from Pakistan (UNHCR, 2002). The reason behind that decision was to address concerns regarding low level fraud and, specifically, to identify 'two-timers' who accessed aid more than once. Refugees would receive their aid entitlement only after they scanned their iris to confirm that there hadn't been an earlier claim against their record. It is impossible to ignore the similarities between the introduction of biometrics in humanitarian settings and the introduction of fingerprinting in India under the British Empire. Just as the colonial authorities introduced fingerprinting because they suspected local people of claiming pensions twice, UNHCR introduced iris scans in order to tackle 'two-timers'. In both cases the introduction of biometrics stems from a relationship of suspicion which sets the tone about the social shaping of technology. Both in the context of the British Empire and in the context of refugee camps, biometric registrations were introduced in very asymmetrical contexts to manage and control populations.

Since 2002 the use of biometrics has been normalized and extended to a range of practices from 'wearable technologies for good', which include examples of biometrically enabled wearable devices to monitor children's health and vaccination programmes (Sandvik, 2020)[4] to cash distributions, which is the most common application. The use of biometric methods was cemented in the 2010 UNHCR 'Policy on Biometrics in Refugee Registration and Verification Processes', which stated that biometrics 'adds value to UNHCR identity, registration and documentation processes by providing reliable identity authentication and preventing risks of false claims, fraud and identity theft' (Office of Internal Oversight Services [OIOS], 2016: 1). In 2015, UNHCR launched a new Biometric Identity Management System (BIMS), in partnership with the global consulting firm Accenture, to capture and store all fingerprints and iris scans from registered refugees (UNHCR, 2015). By the end of 2015, BIMS was rolled out in eleven countries with 593,000 refugee enrolments and a budget of US $9.6 million (OIOS, 2016).

In 2017, UNHCR launched a new data ecosystem, PRIMES, which stands for 'Population Registration and Identity Management Ecosystem' (UNHCR, 2019) which encompasses BIMS as well as other interoperable modules such as ProGres v4, the core UNHCR refugee registration system which records data, but also manages the digital identity aspect of 'beneficiary data' throughout the refugee cycle. PRIMES is part of the 'digital identity and inclusion' policy that has three objectives: (1) empowering refugees through 'web-based economic activities', (2) 'strengthening state capacity', and (3) improving 'the delivery of aid' through 'efficiency gains', which in turn will increase 'client satisfaction' (UNHCR, 2018). By 2021, PRIMES database included 21.7 million individuals, almost half of whom (9.8 million) with biometric records on BIMS.[5] PRIMES is integrated with state registry systems and has varying degrees of integration with other partner tools such as the WFP's SCOPE as well as private service providers such as IrisGuard the biometric technology company that that supports the *Building Blocks* programme, which opened this chapter. ProGres v4 is deployed in 98 countries with 6,900 users, not just UNHCR staff but also partners such as the WFP and government representatives.[6]

The WFP has also intensified the use of biometrics. SCOPE, which is the WFP's 'beneficiary' and transfer management platform, includes biometric data. While UNHCR's PRIMES encompasses different interoperable modules that focus on different areas, SCOPE includes everything: from biometric identity management to assistance management. By 2020 almost 63.8 million identities were registered in SCOPE, with 20.2 million records actively managed through the system.[7] The WFP, which is the largest UN agency and the one most focused on logistics and the distribution of aid, combines biometric technologies with innovations like blockchain to manage cash distributions – increasingly, the preferred form of assistance. One such example is the *Building Blocks* scheme. Given the growing preference for cash assistance instead of aid in-kind, blockchain combined with biometric authentication has been deemed a useful method to streamline and manage transactions. By 2021 SCOPE supported 71 per cent of the WFP's cash operations and was implemented in 68 out of the 85 countries in which the WFP is active (WFP, 2021).

PRIMES and SCOPE represent a clear acceleration of the rate of biometric registrations of refugees. UNHCR announced in 2019 that it aimed to host all refugee data from across the world within PRIMES (UNHCR, 2019) while the WFP reportedly aims to include 80 million biometric records in SCOPE 'in the coming years' (The Engine Room et al., 2023: 19). When such clear targets are set, they often become a bureaucratic goal in their own right, instead of a means to assist refugees. Further, given their size and centralized character, PRIMES and SCOPE are increasingly relied upon by UN partners to provide assistance. The WFP is known to make funding conditional on the use of SCOPE (The Engine Room et al., 2023: 18), which raises immediate questions about whether this contravenes the humanitarian principle of humanity, according to which aid should be directed to all people unconditionally. We can see that these vast databases become the infrastructure that underpins a multitude of humanitarian operations and processes. The infrastructural nature of biometrics raises several questions regarding safeguards as well as risks relating to and the reusability of the datasets. The absence of a dedicated biometric data policy by UNHCR and the WFP compounds such risks which will be discussed later in the chapter. Successive internal WFP audits in 2018 and 2021 found SCOPE 'in need of improvement' to reduce risks in the system for data subjects.[8]

While UNHCR and WFP are accelerating the rate of biometric registrations and expanding the uses of the technology, other humanitarian organizations are more cautious. The International Committee for the Red Cross (ICRC) has the most comprehensive policy on biometrics in the sector published in 2019.[9] The biometrics policy supplements the ICRC Rules on Personal Data Protection.[10] The ICRC policy identifies very specific purposes that call for the processing of biometric data such as identifying human remains recovered from disasters or conflict zones or the restoration of family links. Crucially, the policy is clear that biometric data must never be a condition of service provision, while data-sharing with governments can only occur under the strictest conditions and only with the explicit consent of data subjects. The ICRC policy mandates the deletion of data after the expiration of the retention period. Following a systematic review, Oxfam issued a moratorium on biometrics in 2015. Oxfam adopted a new policy on biometrics in 2021 which specifies that it will only participate in biometrics-based projects if there is a demonstrable benefit to affected people and when data subjects are given a choice about whether to give their foundational data.[11] The policies of ICRC and Oxfam, the only two organizations with specific biometric data policy documents, reveal that the humanitarian sector isn't monolithic and that there are genuine concerns and debates surrounding the risks of these technologies. Yet, the UN, because of its size and because it is the main organization dealing with refugees, has set the agenda.

Biometric technologies, of course, are also widely used by states in the management of migration and refugee flows. As we saw in Chapter 1, biometrics is the *par excellence* technology of the border. Eurodac is the European Asylum Dactyloscopy Database which is shared by the Schengen zone European Union countries to manage asylum applications within the EU. Eurodac is one of the largest biometric databases globally, with 1,481,815 fingerprint records transmitted by member states in 2022 and over 9.35 million records transmitted between 2015 and 2022.[12] Eurodac supports the controversial Dublin Regulation which determines the country responsible for asylum application. When a refugee enters an EU country and applies for asylum, they submit their fingerprints which are stored in Eurodac. If the fingerprints are subsequently resubmitted to a different country then the refugee may be returned to the original country of entry (for detailed discussion see Dijstelbloem, 2021; Metcalfe and Dencik, 2019 among others). Even though Eurodac and border securitization do not fall under the remit of humanitarian organizations, they are key in shaping refugee experiences.

Refugees are likely to come into contact with humanitarian organizations as well as state authorities during their journeys. Further, as we'll see later in the chapter, data sharing between states and humanitarian organizations blurs the boundaries between regimes of data governance. An added complication arises from the role of the private sector, and in particular the biometric-industrial complex, in the collection and handling of refugee data. Eurodac-sanctioned fingerprint scanners used in Greek hotspots 'are produced by a multinational corporation based in Florida and use FBI-compliant software' (Pelizza, 2020: 283). These examples demonstrate why we need an infrastructural approach to biometrics. Such an approach allows us to observe how biometric data travel within, but also beyond the humanitarian field – and the vulnerabilities this creates for data subjects.

In the next paragraphs I will focus on the risks and harms associated with biometric infrastructures.

Bias

During the first ever biometric pilot in humanitarian operations in 2002, UNHCR identified, and turned away, more than 396,000 Afghans out of a total 1.8 million refugees (UNHCR, 2002). Those turned away were deemed to be 'recyclers', a term used to refer to those who claim aid twice. If a refugee could not be authenticated, or if the system identified them as having already claimed their entitlement, then they were turned away. A UNHCR representative in that mission declared his trust in iris technology when he stated: 'How can [refugees] argue now, *the machine can't make a mistake.*'

Let's examine this claim more closely. In 2002, the error rate in iris identification for such a large database would have been 2 to 3 per cent (Bowyer et al., 2008). This suggests that up to 11,800 claimants out of the 396,000 recyclers may have been erroneously denied aid. As Katja Jacobsen (2015: 64) points out, no UNHCR report refers to the risks of false matches and the fact that error rates increase with the size of the database. Developments in technology and computation mean that error rates have decreased. However, biometric technologies are far from failsafe. Technological convergence sometimes heightens the risks associated with each constituent technology. For example, blockchain is routinely used together with biometrics in cash distributions as is the case with the *Building Blocks* programme. Although blockchain is mainly used for verification (rather than identification), which lessens the degree of algorithmic bias, its indelibility can accentuate any erroneous records as data are immutable once entered on blockchain. This can have devastating consequences for the individual concerned as their claims to aid, asylum, family reunification, and safety depend on their biometric records.

Although biometrics are celebrated as the perfect identification technologies, there is ample evidence about their limitations. Studies have pointed out failures when measuring body parts (biometric enrolment) and when processing and matching records. The conditions under which enrolment or matching takes place can affect the reliability of the biometric record. For example, measurements taken in hot, humid or dusty conditions may affect the accuracy of the records. There are particular concerns regarding the reliability of fingerprints as biometric data. Elderly people, manual workers, and those working in the care, health, or beauty sectors are reported to have faint fingerprints – the latter due to manual labour or the handling of chemicals (Nanavati et al., 2002). Iris scans may be considered more reliable but they, too, are affected by age and other factors (Bowyer and Burge, 2016; Hollingsworth et al., 2008). Because face recognition algorithms are trained on datasets overwhelmingly composed of fair skinned subjects they discriminate against darker skinned and female subjects (Buolamwini and Gebru, 2018). Biometric errors occur not just in the enrolment and processing of refugee data, but also at the level of matching biometric records. Because neural networks run on algorithms trained on data which contain human biases (Caliskan et al., 2017), biometric identifications reproduce and therefore legitimate racial, gendered and other forms of discrimination. The probability of erroneous matches increases with large samples (Jacobsen, 2015: 64).

The above examples show that there is discrimination in technology design (Monahan, 2010). Just like film was optimized to capture white faces (Dyer, 1997), biometric technologies also privilege whiteness. Magnet observes that iris scanners or face recognition technologies work better on blue-eyed white males, 'neither too young nor too old, with good eyesight and no disabilities' (2011: 31). Despite the assumption that biometrics are impartial and scientific, biometric data codify existing forms of discrimination (Magnet, 2011), whereas the discourse of science masks racist, sexist, and classist practices. Race, gender, ethnicity, class, disability and age are produced through biometric technologies. The term 'infrastructural whiteness' aims to capture the failure of biometrics to capture a subject's image because their physiological characteristics do not conform to the predetermined classifications which privilege white bodies (Pugliese, 2010: 56). Such technological failures 'to enrol' are never simply a technological matter. They have deeper consequences including the production of race. W. E. B. Du Bois wrote in *The Souls of Black Folk* about the development of a 'double consciousness', which results from 'always looking at oneself through the eyes of others, of measuring one's soul by the tape of a world that looks on in amused contempt and pity' (1903: 8). For Frantz Fanon, blackness is produced by the white gaze (1952). Reflecting on his own experience of being 'crushed into objecthood' under the violent gaze of the other, Fanon develops the notion of the 'epidermal racial schema', which is the embodiment of racial – and colonial – oppression (1952: 89–119). A similar process takes place when the biometric reader fixes a (white) gaze upon the subject.

Linked to the idea of the epidermal racial schema is the notion of 'epidermalization', which refers to the internalization of inferiority as a result of racism (Fanon, 1952). Simone Browne (2015) argues that the prototypical whiteness underpinning biometrics constitutes a form of 'digital epidermalization': an imposition of race on the body through digital means. For Fanon, epidermalization is a deeply depersonalizing and dehumanizing experience which epitomizes the violence of the colonizer (1952). Although bias is not inherent to refugee biometrics, because these technologies are routinely deployed to identify 'suspect' bodies, 'the impact of technological failure manifests itself most consistently in othered communities' (Magnet, 2011: 50).

We can see that contemporary biometrics are the next step in a long lineage of technologies based on fixed and problematic understandings of race and gender. What anthropometry, phrenology, bertillonage and biometrics have in common is the desire to read deviance from body characteristics. The starting point of all these identification methods is suspicion. What connects the Indian people subjected to fingerprinting under British colonial rule, or Afghan refugees under the care of UNHCR is the suspicion that they are trying to game the system. When the motivation behind identification systems is suspicion, that accentuates the bias in technology design.

Technological bias and digital epidermalization are further compounded by the power asymmetries involved in biometric registrations. It is no coincidence that the body which is being measured is typically that of a racialized subject, while the one measuring is typically a white humanitarian. Biometric technologies reproduce relationships of inequity between Western 'saviours' and the suffering former colonial subjects thus attesting to the tenacity of colonialism.

Safeguarding risks

In July 2019, hackers broke into dozens of UN servers triggering what officials called a 'major meltdown'. In January 2022, the ICRC servers hosting data of more than 51,000 people worldwide were hacked.[13] In December 2017, the cloud server of eleven humanitarian agencies was hacked, potentially compromising the personal data of tens of thousands of crisis-affected people (Raymond et al., 2017). These are just some of the recent data breaches concerning humanitarian organizations that have been publicized.

Like all data systems biometric databases are vulnerable to data breaches. In most cases breaches happen when hackers identify and exploit an unpatched vulnerability in the system. The difference with refugee biometric data is the vulnerability of the data subjects. If databases are hacked and the records fall into the wrong hands the consequences can be devastating. This is not just hypothetical: when the Taliban took control of Afghanistan in 2021, they seized the biometric databases that Western donor governments left behind in the rushed evacuation. The databases contained the biometric and other sensitive data of Afghan employees for Western governments and institutions such as the World Bank, potentially endangering them under the hostile regime (Jacobsen, 2022).[14] While this breach did not involve humanitarian organizations, the danger is clear. Displaced people are already at risk; a data breach could potentially contribute to further discrimination, prosecution, or involuntary repatriation. As we have just observed, humanitarian data breaches are not uncommon, although they are not always publicized.

For example, the 2019 UN incident was not reported by the UN until January 2020, a decision which was heavily criticized as displaying lack of accountability.[15] Because of its diplomatic status the UN has immunity from every form of legal process, and it is 'under no obligation to report breaches to a regulator or the public' nor is it subject to Freedom of Information requests (Parker, 2020).[16]

Because biometric data are immutable, the risks to data subjects are immense in case of a data breach. While it may be possible to change someone's password, it is not possible to change their iris. Given the vulnerability of refugees, there are significant potential harms, if their biometric data are leaked. Storing massive volumes of data in universal databases poses significant risks. The remote storage enabled by cloud computing compounds the weaknesses of biometric systems. While no biometric data were compromised in the above-mentioned breaches, it is possible to imagine that BIMS or SCOPE could be potential future targets. This potential vulnerability is at odds with the humanitarian imperative 'do no harm'. The issue is compounded by a lack of policy on biometric data security among humanitarian organizations. Of the six major humanitarian organizations who use biometrics, only two (ICRC and Oxfam) have dedicated biometrics policies.

Further, there is the not insignificant dimension of human errors exacerbated by the lack of resources and data protection training in local offices (The Engine Room et al., 2023). An internal UN audit report identified serious breaches (e.g., leaving workstations unsupervised while publicly accessible) in the deployment of biometric registrations across five countries in 2016, which could 'lead to the loss or misuse of personal data of persons of concern' (OIOS, 2016: 9). Several of my interviewees from the humanitarian sector expressed concerns about the lack of safeguards they had encountered. Interlocutors were careful not to blame local offices who work under extremely difficult circumstances and who are often overstretched. Still, one of my interlocutors saw future scandals as inevitable because there is 'so much bad practice with data; so much unnecessary collection and use without consent and clear ethical procedures'.

Despite conversations across the sector regarding responsible data practices, the absence of clear policy on data practices and data security among humanitarian stakeholders is conspicuous, if not alarming. The foundational nature of biometric data means that the imperative to protect people of concern should also extend to their data. Protecting people's data must be considered a humanitarian priority and this necessitates clear policies on data collection, management, sharing and deleting.

The lack of biometric data policy and standards of practice are also reflected in wider issues regarding data sharing with governments and commercial partners.

Data reusability and function creep

In January 2021 it was revealed that the government of Bangladesh shared the biometric data of Rohingya refugees with Myanmar.[17] There are many examples of biometric data sharing with governments although such practices are usually surrounded by opacity (Jacobsen, 2015 and 2022). A key attraction of digital data is their replicability, retrievability and reusability. Biometric records can be easily retrieved, shared and reused. The very features that make biometrics attractive to humanitarian agencies, donors and private companies also constitute the weaknesses of such systems. Given the vulnerability of refugees and the fragile political contexts in which biometric registrations take place, the risk of data misuse can have grave consequences. This was evident in the case of data breaches, but it is also a problem in the case of data sharing with governments and other partners.

Humanitarian organizations routinely share data with states under their cooperation agreements. All agencies, including UNHCR, operate under the jurisdiction of host nations, which put them under pressure to comply with data sharing requests (Jacobsen, 2015). An internal UN audit report reveals not only the routine nature of data sharing, but also an astonishing lack of safeguards. For example, not only did UNHCR missions share the personal data of refugees with the governments of the Central African Republic, India and Thailand, they did not 'assess the level of data protection applied by the respective governments', nor did they obtain 'transfer agreements' (OIOS, 2016: 11). Once data are shared, UNHCR or other agencies have no power over how the data are stored or used in the future under different governments. Biometric data sharing can facilitate surveillance and function creep, whereby the original purpose of data collection is different from subsequent uses (Ajana, 2013). Such concerns are heightened by the increasing interoperability of databases (Ajana, 2013). The absence of legal frameworks for data and privacy further compounds these risks.

Data sharing does not just occur with governments; private companies, which routinely conduct registrations since 2002, may also have access to data. The agreements between UN agencies and private companies are not publicly available which precludes any meaningful accountability. UNHCR (2019) states that partners can access the PRIMES database, without details about what access is granted to commercial partners and contractors including those that provide the software or hardware for biometric measurements.

On the announcement of its US $45 million partnership with Palantir in 2019, the WFP issued a statement 'that no access to data that provide beneficiary participation would be granted', but did not mention access to metadata, which are equally sensitive and can have deleterious consequences if they end up in the wrong hands (International Committee of the Red Cross and Privacy International, 2018). Two of my interlocutors from the 'digital humanitarianism study' who are involved in private–public partnerships acknowledged that they were aware that data sharing either took place, or that it was technically possible, at least in the context the partnerships they were familiar with. Given the lack of public policy regarding data-sharing practices, opacity reigns. As information infrastructures become vast and interoperable with those of other systems, Mark Andrejevic and Kelly Gates astutely observe that 'function creep is not ancillary to the data collection, it is built into it – the function is the creep' (2014: 189).

Lack of meaningful consent

In June 2019 the United Nation's World Food Programme (WFP) temporarily suspended the distribution of food aid in Yemen when Houthi leaders, representing one of the sides involved in the protracted civil war, opposed the use of biometric data in aid delivery. The WFP, which insisted on biometric registrations as part of efforts to address fraud in aid operations, was criticized for its decision to deny food to a population facing extreme precarity and hunger. Without wanting to enter the details of the complex Yemen conflict here, this episode crystallized the lack of meaningful consent in the biometric registrations of humanitarian subjects.

It is useful to begin our discussion with the definition of consent according to one of the most comprehensive legislations regarding data protection, the General Data Protection Regulation of the European Union (GDPR). According to GDPR, consent must be a 'freely given, specific, informed and unambiguous indication of the data subject's wishes'.[18] As we shall see below, neither of these requirements is fulfilled by current practices in humanitarian settings.

Technically, informed consent is part of the process of biometric registrations in humanitarian settings. UN agencies such as UNHCR and the WFP are expected to provide refugees with information regarding how their data will be used, stored and accessed, especially if data will be accessed by third parties. Informed consent processes must make clear any potential risks before data collection takes place. Crucially, informed consent is based on the assumption that people can refuse to participate without detriment once they have full knowledge of the process and the potential risks. The problem with informed consent in biometric refugee registrations is that opting out is not an option as this would entail loss of aid. As one of my interlocutors from the 'digital humanitarianism study' put it: 'If you don't register with us you are out of our protection here. It is in your interest to give your biometric data.' Refusing to give one's biometric data amounts to refusing aid, when there are no alternatives for food or shelter. Informed consent is rendered meaningless when it is not possible to say no. Without choice, 'informed consent' looks more like coercion.

Apart from having a real choice and being able to refuse, informed consent also depends on the availability of full and clear information regarding the process of data collection and data management. There is plenty of evidence to suggest that this condition is not met. An internal UN audit report revealed that the level of information provided to refugees was inadequate (OIOS, 2016). The report was particularly damning regarding the extent to which refugees had not been informed about their use of data by governments or third parties such as the vendor companies to whom registrations are outsourced. This contradicts the requirement for consent to be specific regarding data processing and access, giving data subjects the opportunity to consent to each use of their data.

> In four out of the five country operations reviewed, OIOS observed that the level of information provided to persons of concern during the biometric registration was below the standards required by [UNHCR] policy. There were also inconsistencies in the information provided, particularly regarding the access to the data by third parties. [...] There was no evidence that persons of concern were informed of their rights and obligations [...]. (OIOS, 2016: 10)

This criticism is echoed by research reports, one of which found that refugees in Za'atari were neither aware of how *Building Blocks* worked, nor were they told what data were collected about them and who had access to it (Shoemaker, Currion and Bon, 2018: 19). Having a basic of understanding of how an identity system works is a fundamental prerequisite for meaningful consent. It is impossible for anyone to give consent to something that they do not understand. As Andrew McStay notes: 'without [information] there is no consent, but rather the application of force [...] To be devoid of understanding is to be unable to give proper consent' (2013: 600).

Other reports from Za'atari reveal that refugees had not even been asked to consent to receive their monthly allowance via the *Building Blocks* pilot.[19] One theme that emerged from my interviews in the humanitarian sector is the complexity of the technological systems – which need to be explained – and the need to streamline and expedite registrations and cash disbursement given the finite resources of aid agencies. Invariably, the latter prevails over the former. The lack of understanding of how biometric and blockchain systems work and how personal data will be used is not just a matter of ethical practice, but, crucially, also a matter of trust. It is not surprising that refugees in Za'atari express misgivings about the digital identity platforms. These concerns are often expressed in embodied ways as we will see in the following section.

Some of these critiques reverberate the discussions around digital consent more broadly. Consent is a contested concept, even in less asymmetrical settings such as the ones explored here. For Solon Barocas and Helen Nissenbaum consent is deception: 'consent can never fully specify the terms of interaction between data collectors and data subjects' (Barocas and Nissenbaum, 2014: 45). Elinor Carmi has argued that digital consent has been used to 'authorize and legitimate exploitative and harmful practices to make the [digital capitalism] business model work' (Carmi, 2021: 8–9). Ultimately, digital consent normalizes surveillance and the commodification of human data. All these concerns are magnified in humanitarian contexts which involve some of the most sensitive data of some of the world's most at-risk populations.

It is evident that informed consent is rendered meaningless in highly asymmetrical settings such as refugee camps. When biometric registration becomes the condition for refugees to receive aid and shelter, then consent does not carry any meaning. When aid is only distributed via biometric authentication through blockchain infrastructure then opting out is impossible. In situations of extreme dependency and when power relations are so unequal, consent will always be contested.

Consent originally aims to address power inequalities by informing people and enabling them to exercise their agency. By contrast, on this occasion, consent further entrenches and legitimates the asymmetrical relations between aid receivers and aid providers: humanitarian organizations, governments and private companies. Because consent provides an illusion of agency, the problematic uses of biometric technologies are legitimated. This, perhaps, is where the greatest problem lies. Consent in humanitarian settings normalizes and legitimates a system of dependency, control, and, as I argue in the section below, coloniality.

Coloniality of digital identity programmes

Digital identity programmes are championed as emancipatory by enabling refugees to be in control of, and have ownership of, their data. The phrase 'digital wallet' is typically used to refer to blockchain and biometrically enabled cash assistance such as *Building Blocks*. The term 'digital wallet' conjures up the image of the financially independent refugee, who can save, shop freely or even become an entrepreneur. The reality in the refugee camps in Jordan and Bangladesh could not be more different.

The *Building Blocks* scheme is a cashless system, but it severely restricts where the money can be spent. Just to take the example of Za'atari, cash assistance can only be spent in the two designated supermarkets and the four bakeries located inside the camp.[20] *Building Blocks* is cashless and virtual, but its transactions are geographically confined to six stores. The 'digital wallet' does not even extend to the camp as a whole and the informal cash-based market of *Shams-Elysees* where local people repair, sell or exchange goods. One of my interlocutors in the 'digital humanitarianism study' with experience in the camp, confirmed that refugees preferred the possibility of receiving cash and spending it in the outlets of their choice. Having greater choice than the two UN-sanctioned grocery stores, matters even more in the context of high inflation and rising prices following the Covid-19 pandemic and the war in Ukraine. Goods can be cheaper in the *Shams-Elysees* street market, which runs along the main street of the camp, but refugees are not allowed to use their 'digital wallets' there. A recent report found that supermarket prices in the Za'atari camp were significantly higher compared to supermarket prices elsewhere in the country. A 2022 report compared grocery prices in the Jordanian capital Amman branch of the Safeway supermarket and the Za'atari branch of Safeway supermarket, which runs the *Building Blocks* scheme. While a litre of milk cost 2.99 Jordanian dinars in Safeway Amman, it cost 3.25 in the Za'atari branch.[21] When refugees receive 23 dinars a month, such a difference in price is very significant.

The 'digital wallet' is not emancipatory enough to extend to the most entrepreneurial activities of the camp along the *Shams-Elysees* street. Named after *Champs-Elysees*, the famous shopping avenue in Paris and the world *Al Sham* which is an informal name for Syria and the surrounding region, the local marketplace includes barbers, repair shops and vendors selling sim cards, clothes, sweets among other things. It is not possible to manage one's finances, save or transfer money through the 'digital wallet'. The combination of biometrics and blockchain controls where refugees can spend their money, and ultimately, restricts the movement of refugees.[22] Cash assistance is promoted as a response to critiques about the paternalism of 'aid-in-kind'. But if refugees can only shop in limited outlets with a fixed range of options, biometric and blockchain-enabled cash essentially reintroduce a form of 'aid-in-kind'. The 'digital wallet' is meant to empower, but its functionality is to control.

It is hardly surprising then that Za'atari residents were unhappy about *Building Blocks*. Margie Cheesman in her ethnography of Za'atari observes that virtual money clashed with local idioms of exchange and reciprocity: the local preference for the tangibility of cash was in sharp contrast to the virtuality of blockchain transfers (Cheesman, 2022). Money, of course, is much more than just a financial transaction. Money reflects relationships and moral orders specific to empirical contexts (Bloch and Parry, 1989; Maurer, 2006; Zelizer, 1997). Imposing an imported model of virtual transactions constitutes a form of epistemic violence – a feature of colonialism and coloniality. It's no wonder that refugees in Za'atari have expressed strong reservations about the biometric scans. The strongest objections focused on the potential harms to the body. The iris scanner was experienced as intrusive, while residents have expressed concerns about the impact of radiation. According to an investigative report, refugees have even complained of pain following biometric scans.[23] These expressions of embodied discomfort make complete sense given the top-down imposition of these systems and the opacity that surrounds them. The same Redfish report found that many refugees had not even been asked whether they were happy to participate in the *Building Blocks* pilot. We can see here how the lack of meaningful consent in these highly asymmetrical situations further accentuates the problematic nature of digital identity infrastructures.

The bodily discomfort reported by refugees in response to biometric scans echoes the comments of the Indian labourers in the nineteenth century Transvaal when they were subjected to fingerprinting by the British colonial authorities. Recall that in that context, people experienced fingerprinting 'like a dog collar put on us' (Singha, 2000: 194). Biometrics and blockchain may be championed as contributing to humanitarian reform, but they smack of paternalism and coloniality. In the *Building Blocks* example, we see that the infrastructures of digital identity become the ambient background of everyday life and, in so doing, produce subjectivities.

Conclusion

The chapter began by tracing the genealogies of biometrics back to the nineteenth-century practices of anthropometry and bertillonage and the genealogies of the British Empire. The historical perspective revealed continuities with the contemporary application of biometrics in refugee camps to manage and control populations. Another pattern involves the endurance of suspicion which animates the design biometric systems, whether in colonial India or in contemporary refugee camps. It is not surprising that refugees, just like colonial subjects over 100 years ago, experience biometrics as deeply invasive. Despite technological innovation, biometric methods continue to be based on problematic classifications regarding race, gender, class, age and disability and as a result discriminate accordingly. In so doing biometrics racialize and marginalize populations.

What emerged clearly in this chapter is the infrastructural character of biometrics, which increasingly underpins other systems and becomes ubiquitous and interoperable. *Building Blocks,* the virtual cash assistance project, exemplifies the infrastructural character of biometric systems which combine with blockchain and commercial infrastructures to disburse payments. These infrastructures allow humanitarianism to extend into new spheres of governance, and other forms of governance – such as the state and private companies – to extend into humanitarianism.

The chapter argued that biometric practices constitute a form of structural violence. The introduction of biometrics in camps is a form of epistemic violence, the imposition of a Eurocentric system that clashes with local idioms about money and respectability. Biometrics reduces identity to a transactional category while stripping people of their agency to define themselves. Biometric technologies are also technologies of control and can even include physical violence as we observed in the Rohingya example in the Introduction chapter. The Rohingya example illustrates the collusion between the state, private companies and the humanitarian sector.

Concerns regarding privacy, safeguarding and function creep add to the potential harms. The lack of meaningful consent in refugee biometric registrations further compounds some of the above inequalities. While it is theoretically possible for a refugee to refuse biometric data collection, this is not an option for most refugees as that would amount to refusing aid when no other livelihood options are available. Without meaningful alternatives and without full knowledge of how biometric and blockchains systems work there can be no freely given consent. Ultimately, digital identity practices reconfirm the hierarchy between aid providers and refugees – and in so doing reaffirm that, structurally, contemporary versions of humanitarianism are not dissimilar to their colonial counterparts. Biometric technologies in humanitarian settings exemplify technocolonialism. They rework the colonial genealogies both of humanitarianism and technology and resurface them in tangible ways: they contain the agency of refugees and produce race and subjectivities by imposing a white gaze on racialized and othered bodies.

Even if it is meaningless, consent performs an important role. It occludes the power relations and legitimates and normalizes some of these practices. Similarly, the discourse of empowerment (exemplified in the fantasy of 'digital wallets') and the discourse of progress mask the inequities in the system. In the next chapter we'll explore how accountability performs a similar role in occluding power asymmetries.

3

Extracting Data and the Illusion of Accountability

In the months following the landfall of Typhoon Haiyan, one type of poster was ubiquitous in the outskirts of Tacloban and the island of Sabay. It could be found pinned to the tents that housed people displaced by the Typhoon, on the trunks of trees and by the *sari-sari* stores. Each poster differed depending on the aid agency which produced it, but all posters included an invitation to submit feedback to a dedicated hotline. The poster included the relevant number where people could send an SMS with their comments for free. The poster specified the humanitarian agency concerned as well as the particular intervention for which feedback was sought. A typical example would be: 'agency x seeks your views on the water and sanitation initiatives in your *barangay* (neighbourhood). Please send an SMS to this number for free.'

The feedback hotlines were part of the 'accountability to affected people' (AAP) programmes that have become a mandatory component of aid projects. The interactive nature of digital technologies is assumed to enable affected people to participate in the recovery process and hold humanitarian agencies into account. The high level of social media and mobile phone use in the Philippines were seen as opportunities to digitize AAP initiatives. As discussions on the need for greater accountability to affected people were intensifying in humanitarian sector, the response to Typhoon Haiyan provided 'the perfect laboratory' to implement accountability programmes using digital technologies, as several participants put it.

Of all the ways in which accountability can be assessed, feedback mechanisms have been prioritized as the most relevant. This is a narrow definition of accountability which mirrors the way accountability is measured in public life more broadly.[1] Accountability, of course, can be evaluated in more complex ways. For example, the participation of affected communities and the degree of engagement in the recovery process can be valuable indicators of accountability, albeit rather challenging to quantify. The advantage of feedback surveys is that they simplify and streamline the way affected communities respond to aid operations. If feedback is a narrow definition of accountability, it is rendered even narrower in humanitarian settings where it is framed as feedback in relation to specific interventions. Agencies typically solicit views in relation to a particular project they deliver – not the overall response. For example, in the above example, agencies asked for feedback on a particular water and sanitation (WASH) project, not other aspects of the response such as food or cash distributions, shelter or livelihoods'.[2] Further, feedback is further narrowed down as 'digital feedback'. While feedback can be collected through various methods (community consultations, informal conversations, letters in drop boxes, comments on social media, and in-person agency visits to name a few), increasingly the feedback that finds its way in the official reporting channels is the content from digital sources: SMS hotlines, and more recently, messaging services, bespoke feedback apps and chatbots.

One warm and humid afternoon in June 2014, Jonathan Ong, Liezel Longboan and I arrived in the village of Maya in the north of the island of Sabay. We were keen to spend some time outside the main town, where we were mostly based, in order to hear about people's experiences in the countryside. Aid was taking longer to reach the villages in the north of the island where fisherfolk were particularly affected having lost their boats during the Typhoon. We arrived in the village of Maya in the early afternoon and headed towards *barangay* Kawit. This was one of the poorest neighbourhoods where most houses had been significantly damaged during the storm. By the time of our visit the roads had been cleared, although debris and fallen trees littered the roadside. Some houses had new iron roofs installed and new pots with plants and flowers were beginning to bloom in their front yards. A makeshift hairdresser was doing brisk business by the roadside. Dolores' house looked derelict even compared to those of her neighbours. As we walked towards her house Dolores came out to greet us. It soon transpired that she assumed we were aid workers finally responding to her feedback message which she sent in February 2014 – four months before our visit.

Dolores is mother to two children and is married to Jeri, a low-income *trisikad* (nonmotorized tricycle) driver. Their house was severely damaged by the Typhoon. An old tarpaulin covered the roof, which leaked during the rainy season. Dolores was disappointed because aid agencies in Sabay had prioritized those working in the fishing sector. The island, well known for its fishing, received several targeted interventions to fisherfolk to the almost total exclusion of other sectors, such as transport. As a result, Jeri and his family received no livelihood assistance. Dolores told us how she used the SMS feedback hotline in order to express her concerns and request assistance. She only received an automated acknowledgement from the agency with no follow-up response. When we met her four months after she had sent the text, Dolores still hoped that the agency workers would respond so that she could explain her family's predicament without aid. Dolores had turned to relatives, friends, and even lending agencies to provide for her family and carry out the most basic house repairs. She still wished for an opportunity to express the suffering that she had not been able to convey in a laconic SMS: 'My name is Dolores. Please also help the *trisikad* drivers. I live in *barangay* Kawit.' Dolores' experience of just receiving an automated acknowledgement was typical of half of our interlocutors who used the formal feedback mechanisms.[3]

If Dolores and others did not receive a response to their feedback, what was the purpose of the 'accountability to affected people' programmes? Who was the recipient of these feedback messages and what happened to the feedback data? How can the deafening silence to Dolores' grievances be compatible with the 'most accountable response to date'? To answer those questions our team followed the feedback data trails. This chapter traces the journey of feedback data: from the narrow definition of accountability as 'digitized feedback' and the implementation of the relevant policies, to the collection, aggregation, analysis, processing and reporting of feedback. The discussion will draw on my fieldwork in the aftermath of Typhoon Haiyan as well as additional interviews during the 'digital humanitarianism project'. Even though the feedback collection channels have changed since Haiyan, the patterns remain the same regardless of whether the feedback technology involves messaging apps, bespoke apps or even chatbots.

Feedback data are extracted from affected communities and used for the purpose of evidencing impact which is vital for the renewal of funding by donors. The digitization of feedback facilitates the aggregation of data and in so doing hastens the repurposing of accountability data into audit trails. The trails of digitized feedback data reveal the power geometries of humanitarianism. While tracing the trails of digitized feedback data the chapter will also uncover the harms associated with how 'accountability to affected people' initiatives are practiced. The colonial legacies are underlined by the fact that accountability is essentially a Western notion that does not translate into local cultures. The use of chatbots and AI to streamline accountability reporting further compounds these issues. Chatbots are designed in the global North and are trained in the English language. By including preselected answers these systems already pre-determine what 'feedback' is.

Accountability to affected people as feedback

As we saw in Chapter 1, the interactivity of digital technologies is seen as an opportunity to correct the power asymmetries of humanitarianism by involving people in the recovery and making agencies more accountable. Haiyan was branded as 'the most accountable response to date' because of the available communication infrastructures (Jacobs, 2015). The Philippines has been popularly referred to as the 'social media capital of the world' while mobile teledensity already exceeded 100 per cent in 2013.[4,5] Mobile teledensity refers to the number of mobile phone connections per one hundred inhabitants.

What 'the most accountable relief response to date' translated into is a proliferation of digital feedback platforms. Almost all relief projects had an 'accountability to affected people' (AAP) element which typically involved a feedback mechanism. Most agencies had at least one accountability officer while the larger agencies had whole accountability teams dedicated to processing feedback. The largest AAP team I encountered numbered fourteen people who were tasked with coding the messages, aggregating the feedback messages and preparing reports. Haiyan was the first time that an AAP Interagency Coordinator was deployed (Wigley, 2015). Accountability programmes were well resourced even in what were challenging circumstances for NGO staff.

Following the 2016 World Humanitarian Summit and 'Grand Bargain' agreement, which advocated a 'participation revolution', accountability to affected people efforts have further intensified. Since the response to Typhoon Haiyan, we have seen the introduction of a range of new platforms, from messaging apps to chatbots that are used to collect feedback. The mix of feedback methods varies according to context, but digital methods seem to have become dominant. What hasn't changed since Typhoon Haiyan is the narrow definition of accountability as feedback. As we observed in Chapter 1, of all the possible ways in which accountability can be defined and implemented, humanitarian agencies have consistently chosen to focus on feedback in every single operation since the days of the Humanitarian Accountability Partnership (HAP) in 2007 (Krause 2014).

The narrowness of accountability as feedback is further compounded by the fact that agencies solicit feedback only in relation to specific interventions. Feedback only counts if it relates to actual aid projects. If a humanitarian agency focuses on a sanitation project, it only processes feedback in relation to the sanitation project. Broader concerns, including the recovery as a whole, do not count as feedback. The compartmentalization of feedback might be understandable from the point of view of humanitarian organizations, which are responsible for delivering a specific project. But from the point of view of affected communities, compartmentalization does not make sense. Our interlocutors in Tacloban and Sabay were concerned about the wider conditions of their daily lives, not isolated initiatives. The following examples illustrate this.

One of the most strongly felt grievances expressed during our fieldwork concerned the selective nature of aid distributions. Given their limited resources, aid agencies have to make decisions about those deserving assistance and those who do not. In the Haiyan response, aid focused on specific sectors or individuals who were deemed particularly vulnerable. Only aid recipients were expected to submit feedback. Unsurprisingly, those excluded from aid programmes often felt very strongly about their exclusion. They were the ones who were keen to express their grievances as they had something to complain about – their exclusion. Such was the case of Dolores and Jeri whom we met at the beginning of this chapter. Jeri, like most working in the transport sector was excluded from the distribution list. Dolores' used the feedback hotline to draw attention to her case and urge the INGO to help her family as well. As we saw in the chapter's introduction, the feedback of those excluded doesn't count as feedback.

Similar stories can be found across empirical contexts. In the 'digital humanitarianism project', an interlocutor from the humanitarian sector who worked in the Middle East shared with me that the majority of feedback messages in their agency hotline were from those excluded from cash assistance. Yet, the feedback of those excluded from the relief efforts simply doesn't count as feedback. Accountability officers are only expected to respond to feedback that pertains to the delivery of specific programmes. As a result, they are not able to act on feedback that extends beyond the agency's narrow remit.

The selective nature of aid distributions reveals a fundamental problem with humanitarian assistance. Humanitarian agencies often rely on local governments to guide them with the criteria that will help them decide eligibility, although as we shall see in Chapter 5 such decisions are increasingly processed algorithmically. In the Typhoon Haiyan response, *barangay* leaders decided that aid will be distributed to 'heads of households'. By household heads they meant married males. This meant that unmarried mothers were excluded from the distribution. This was the case of one of our interlocutors, Lia, a single mother of two boys, who fought daily to be recognized as a household head in order to receive assistance to feed her family. Her efforts were in vain. Lia's complaints did not count as feedback as she was not eligible for aid. The humanitarian agency in question could not bypass local leaders, who in turn were not prepared to change their ways. In this example, by accepting existing social hierarchies and politics, humanitarian agencies reproduced inequalities and exclusions. In Lia's case the selective distribution of humanitarian aid institutionalized patriarchy and heterosexual marriage. It is worth nothing that in the Philippines, patriarchal norms have colonial roots. The period of Spanish colonialism altered the status of women, who in the pre-colonial period, evidence suggests, were known to hold positions of religious and political power (Torres, 1987: 312; see also Lugones, 2007). The Roman Catholic Church has been instrumental is shaping gender ideologies and controlling women's sexual and reproductive rights (Parreñas, 2001). Divorce continues to be illegal in the Philippines, as is abortion. Colonialism permeates local institutions and local leaders who have internalized its tenets.

We observe yet another example of the entanglement of humanitarianism and colonialism. The feedback system, which was introduced precisely to address such critiques and empower local communities to hold humanitarian agencies to account, appears here to have the opposite effect. By excluding sections of the population – in Lia's case because of her gender and family status – feedback channels reproduce uneven relationships some of which are entangled with colonial hierarchies. Rather than helping people to hold humanitarian agencies to account, feedback channels silence those who have grievances because they are left out. In so doing, feedback channels play an active role in the process of exclusion and marginalization. They become part of the problem that they are meant to be addressing.

As we disentangle the relationship between aid, feedback practices and colonialism, it is important to remember that accountability is a Western notion that does not translate well into local contexts. For example, there is no word for accountability in Filipino or other regional dialects. In our Philippine fieldwork, humanitarian workers mentioned that people were not used to expressing complaints. One of our interlocutors who worked as an accountability officer told us that in their database, 90 per cent of the feedback messages were 'thank you' notes by people wanting to express their gratitude for the aid that they received.[6] In our visits to the offices of aid agencies I personally saw several of the 'thank you' notes that had been posted through suggestion boxes displayed on walls and notice boards. The feedback hotline databases we were able to access confirmed the popularity of 'thank you messages' such as the one below:

> Good evening all I can say about the help you gave us is THANK YOU SO MUCH to all of you wish you can continue to help us, may GOD repay you for your assistance.

In our fieldwork on the island of Sabay I was struck by the so-called 'thank you shrines' built by affected people to express their gratitude to aid agencies. These 'shrines' were typically built with pebbles and seashells usually forming the phrase 'thank you' followed by the name of the aid agency. The shrines were found on the side of busy streets, roundabouts and the busy port and materially inscribed gratitude onto the physical environment. We even found 'thank you shrines' on the beach where fishing nets destroyed by the storm had been plaited with shells into a letter formation: 'thank you and God bless'.

The 'thank you shrines' reveal that aid is filtered through idioms of gratitude and the experience of colonial encounters. Aid is perceived as a gift which is offered in an asymmetrical power relationship. Many of our interlocutors found it unthinkable to publicly express complaints about what they perceived as an act of generosity by powerful outsiders. The 'thank you' shrines speak volumes about the silence that accompanies the burden of aid as gift. In this context of asymmetrical power relations and colonial legacies, how can it be expected that feedback mechanisms will deliver accountability to affected people? The 'thank you' shrines and the 'thank you' feedback messages reveal that accountability practices inadvertently become a tool of coloniality. They impose a Western framework on what is already an asymmetrical relation of power. It should not be surprising that the response is one that confirms and entrenches the power asymmetries. The paradox here is that the process introduced to democratize humanitarianism had the opposite effect than the intended one. Ironically, this took place in the self-styled 'most accountable response to date'.

The public expressions of gratitude should not be taken to mean that people in Tacloban and Sabay were not critical of aid operations. The public expression of gratitude existed in parallel with private feelings of resentment and anger as many people struggled to rebuild their homes and mourn the dead relatives. Many of our interlocutors were aggrieved about the agencies' selection criteria regarding who deserved aid. Others complained to us about the chaotic relief distributions. The inequalities caused by the agencies' selective aid distribution caused resentment within tight-knit communities with a strong spirit of *bayanihan* which refers to the sense of civic unity and tradition of cooperation when recovering from crises, causing some of our participants to exclaim: 'neighbours and friends have become enemies because of these NGOs'. I will return to the criticisms towards the relief efforts as a form of resistance in Chapter 6. What is striking, is that the most critical comments were not expressed as 'feedback' in the official channels, but were shared with us in the context of the ethnography. Many critical comments were also posted on social media.

Lorna, a young mother from Tacloban who lost her husband during the storm surge, complained to national media about the agencies' 'beneficiary selection criteria' – a very public expression of grievance, which she hoped would draw attention to her case. Her complaint was not acted upon, as it did not register in the agency feedback databases that accountability officers are tasked to monitor. Feedback relayed in informal channels (such as broadcast or social media) is not recorded, digitized or tracked and does not carry the expectation to 'close the feedback loop'. From the point of view of NGOs, such comments do not count as feedback. However, given the proliferation of media platforms, feedback is increasingly likely to be found outside the formal feedback channels. Many of our interlocutors discussed the slow pace of the recovery on Facebook.

Moreover, NGO feedback excludes comments captured through the national government's own feedback mechanisms. Five of our interlocutors used these platforms, often to complain about the politics of aid distribution (which was largely implemented via local government structures and thus reproduced existing patronage relationships). This illustrates that feedback cannot be neatly disaggregated into government – or aid agency – related, although these are treated separately.

Similar observations were made by several of my interviewees in the 'digital humanitarianism project'. They observed that refugees often make searing comments on social media, but these do not count as feedback as they extend beyond the NGOs remit and their official channels. The following quotes from two humanitarian workers based in Greece during 2015 to 2017 illustrate this:

> People talk about the discrimination they face. For example, Afghans are not eligible for relocation. And so they're very frustrated about that. [...] They say, 'we have bombs in our country, too, we're dying too, why are they not relocating us [like] the Syrians?' So there's a great sense of discrimination that they feel and they need to talk about that a lot. But all this is on social media so we can't officially record it.

> Some refugees would say [on Facebook] that Ramadan distribution times were problematic because of fasting. But there was not a huge amount we could do with that information also because the detention centres are managed by the Greek government.

These quotes reveal a wider problem which relates to the fact that feedback cannot be disaggregated according to the different stakeholders involved. Afghan refugees in Greece, just like the people of Tacloban, face one reality which cannot be compartmentalized into different bureaucratic units.

Automating accountability? Why digitizing and automating feedback is not a solution

We have so far observed the entanglement of aid, accountability and coloniality as well as the problems that arise from the narrow definition of accountability as feedback. Another way in which accountability is further narrowed down, is through the prioritization of *digital* feedback channels. Increasingly, of all possible feedback collection methods, digital platforms, are prioritized as the key channel. This was certainly the case in the Haiyan response where, although international agencies had several feedback mechanisms at their disposal, including face-to-face consultations, suggestion boxes, and helpdesks, feedback obtained through SMS was more likely to be included into the agency feedback logs. For example, one agency's feedback database included over 400 per cent more entries from SMS hotlines compared to community consultation and helpdesk entries.[7] Our humanitarian respondents attributed the problem to humanitarian bureaucracy and the associated division of labour. Frontline staff who are responsible for aid distributions and who are in daily contact with affected communities have first-hand understanding of people's concerns. Yet, our interlocutors told us that when frontline staff came across complaints, they referred people to the SMS hotline. It was clear to frontline workers that it was not their job to relay feedback to the accountability team. Frontline staff felt ambivalent about this. Some would like to be able to report, and ultimately, resolve issues; but many of our interlocutors thought that the SMS feedback hotline was a positive development in that it formalized complaints, therefore making them more visible. According to some of our interviewees the tangible trail of digital feedback can add weight to concerns and track the progress of addressing them. This echoes well-documented bureaucratic practices, which prioritize written complaints (Gupta, 2012).

The prioritization of digitized feedback is illustrated by the experience of one accountability officer who recounted that as soon as she arrived in Tacloban 'the first thing she was asked to do as an absolute priority' was to set up the agency's SMS hotline. Our interlocutor did so, only to realize weeks later that this method did not offer insights into the concerns of the community. Similar stories are echoed in the accounts of my interviewees in the 'digital humanitarianism project'. Many accountability officers observed the disconnect between what they learn via the official 'complaint and feedback mechanisms' (CFM) and the understanding they develop via other sources in the countries in which they are deployed. One interlocutor stressed that none of the serious sexual abuse scandals in the aid sector that have surfaced in recent years were reported via the CFM channels.[8]

The prioritization of digital feedback methods during the Haiyan response was in direct contrast with local residents' preference for in-person consultations compared to technologically mediated ones. Some of our interlocutors felt disappointed when agency staff directed them to the formal feedback mechanisms. While in the Haiyan response there was a clear preference for in-person feedback (Hartmann, Rhoades and Santo, 2014), there are examples that digital feedback platforms may have increased the participation of marginalized populations in other contexts. For example, in in the 'digital humanitarianism project', interviewees from the aid sector with experience in the Middle East and Africa, mentioned that the anonymity afforded by online feedback mechanisms enabled some members of communities to report issues such as gender-based violence and discrimination.

Although there is recognition that feedback methods need to reflect local preferences and norms and combine a range of channels (from in-person to online) the trend since 2014 seems to be the increasing prioritization of digital channels. The proliferation of digital channels from SMS hotlines and messaging apps to bespoke apps and chatbots is not simply a matter of adding further media of communication. Digital platforms are structured in specific ways and therefore facilitate some types of interaction more than others. Understanding the affordances (Baym, 2015; Bucher and Hellmond, 2017) of online feedback platforms helps us make sense of the shape of digitalized accountability. For example, the restriction to 160 characters in text messages can constrain the way a complex problem is expressed. In addition, our analysis needs to take into account the important matter of digital exclusion and safeguarding.

Digital feedback channels typically involve feedback in writing. In so doing they continue a long bureaucratic tradition that devalues oral complaints and prioritizes the written form (Gupta, 2012). Writing has always been the cornerstone of bureaucracy (Goody, 1986). The emphasis on the medium of writing becomes problematic when imposed on marginalized populations who may not be literate. Refugees and people affected by disaster are some of the world's most disadvantaged people. A study of humanitarian feedback processes in Myanmar found that written feedback methods were hardly used, as not everyone could read and write (Hilhorst et al., 2021). Literacy levels among affected communities vary. The average adult literacy in Myanmar is 89.5 per cent,[9] while literacy levels among the Rohingya in Bangladesh are estimated to be between 73 and 93 per cent.[10] As always, such statistics are likely to affect older people and women disproportionately.

Digitized feedback also excludes those without access to a mobile phone. Despite the popularization of mobile phones, our poorest interlocutors in the post-Haiyan fieldwork faced significant mobile phone access issues. The issues included mobile phone ownership, but also connectivity which was often very patchy depending on people's ability to pay for airtime.[11] Those without mobile phone or internet access are the poorest people, who are usually those most adversely affected by disasters (Madianou, 2015). If I encountered these inequalities in the Philippines with its high mobile teledensity (144 per cent in 2022), the inequities of access are even more pronounced in countries with lower mobile phone subscriptions per one hundred inhabitants such as Haiti (64 per cent), the Democratic Republic of Congo (50 per cent) or South Sudan (30 per cent).[12] DRC, Haiti and South Sudan are no strangers to humanitarian emergencies. DRC has experienced six Ebola virus outbreaks since 2018.[13] In 2023 DRC faced armed conflict with over 5.6 million internally displaced people needing emergency assistance.[14] Haiti experienced one of the most catastrophic earthquakes in 2010 and an ongoing, complex humanitarian crisis with almost half of the population (5.5 million) facing acute hunger in 2024.[15] South Sudan is dealing with Africa's largest refugee crisis after a decade of conflict.[16]

An additional problem of SMS feedback is its limited length of 160 characters. In the 'Haiyan project', our (admittedly few) interlocutors who used the SMS feedback method struggled to summarize their complex circumstances in such a short space. Even though the limitation is superseded as agencies have moved to messaging platforms with no character limit, the perception often remains that messages must be succinct. The content permanence and shareability of digital content also raises fears that people might be penalised if their comments are misinterpreted. The most common fear during the Haiyan response was that someone or their family could be 'struck off the distribution list'. We encounter once again the power geometries of humanitarianism and how they shape digital practices. This example also illustrates how the affordances of digital content (content permanence, shareability) accentuate the power asymmetries in the context of relief operations.

Since the Haiyan relief efforts, feedback mechanisms have expanded to other digital platforms, most typically messaging apps such as WhatsApp. While there are advantages of using platforms that may already be popular among affected people – notwithstanding the significant issue of digital exclusion – other problems emerge regarding data safeguards. By using platforms owned by large social media companies like Meta, humanitarian agencies become implicated in the business model of social media companies that extracts user data for profit. There are privacy and safeguarding issues here, which we'll further discuss in Chapter 4 on technological experimentation.

The latest iteration of technologized feedback comes in the form of chatbots. Feedback chatbots typically rely on an AI model using natural language understanding (NLU) and natural language generation (NLG) to autonomously answer questions that are submitted in the form of typed messages on messaging apps. Such chatbots may run on WhatsApp or Telegram. Once questions are received and processed, the AI model then selects a response from a list of pre-determined text responses. Some examples include the *Tawasul* chatbot which was launched in Libya by the WFP-led Emergency Telecommunications Cluster in 2020. *Tawasul* (meaning communication in Arabic) runs on Telegram and responds in Arabic and English.[17] In 2022 a new version of the chatbot was launched in the humanitarian response to the Ukraine. The chatbot, which again runs on Telegram, is named *#vBezpetsi*, which means 'safe spaces' in Ukrainian.[18]

Chapter 4 will examine chatbots as an exemplar of experimentation. We will explore the role of language in machine learning and its pivotal role in asserting the coloniality of power. I want to flag one aspect of automation here, namely the fact that low-level chatbots respond by choosing from a sample of 'pre-determined' responses. We can imagine that this method could work well if the question concerns something simple such as 'where is the nearest hospital'? But as we have already established in the chapter, the problems refugees or disaster affected people face are immensely complex. How can pre-determined responses help respond to complex questions regarding a family's exclusion from a cash assistance programme? How can predetermined responses address the multiple levels of exclusion exclusion of what counts as feedback and digital exclusion? As the feedback platforms become automated, the likelihood of digital exclusion increases.

Even for simple queries, chatbots can be inadequate. I personally experienced this when I interacted with a humanitarian chatbot, the WFP Foodbot which was piloted in the Kakuma refugee camp in (Madianou, 2021). When I asked the chatbot the date for the next cash distribution in Kakuma, the bot gave me the wrong date. Such erroneous information can have significant consequences in situations of vulnerability such as refugee camps. This example also raises the question of accountability – who is responsible in the case of accidental misinformation? Is it the humanitarian agency? The bot programmers? The platform owners?

What I found in my interactions with the Foodbot over several months was that it could not respond to several of my questions. Every time I asked a question that deviated from the pre-scripted menu the whole conversation had to be reset. For example, the Foodbot was unable to respond to my query regarding why refugees in Kakuma receive a higher amount than those in the neighbouring refugee camp of Kolobeyei (for further discussion see Madianou, 2021). The frustration I experienced in my chatbot interactions must be a familiar feeling to anyone who has interacted with automated answering systems, which are very common across government departments, or customer service. In order to elicit a response from the chatbot, I found myself effectively sounding like a chatbot. I phrased my questions to sound chatbot-like so as to be legible by the AI programme. In other words, the chatbot imposed a way of communicating with it (see also Suchman, 2007). It set the tone. If I felt 'disciplined' by the chatbot, it is likely that the intended users may also experience this 'disciplining' (see Greiffenhagen, Xu and Reeves, 2023 on the 'self-disciplining' effect of a similar AI programme). In Chapter 5 we will observe that these AI-driven platforms can be understood as 'epistemic machinery' – not just part of knowledge production, but part of a machinery of knowledge production (Knorr-Cetina, 1999).

These observations raise concerns about whether such platforms can facilitate feedback, accountability and participation, which was the original aim of the 'Grand Bargain'. How will predetermined responses help to achieve 'the participation revolution' is puzzling to say the least. Chatbots crystallize something that was included in previous iterations of feedback platforms, but was not as visible. The emphasis on predetermined responses reveals the essence of digital feedback mechanisms which is to measure what they have set out to measure. In other words, feedback is written before it is even collected. This is typical of all bureaucracies as we will see in Chapter 5. Chatbots provide the perfect technology for what feedback systems have always tried to measure: a very narrow response to an isolated activity. We have seen that each technological iteration of feedback further narrows this limited remit. What the discussion has revealed so far is that digitized feedback – because of its narrow definition and modalities – is a far cry from participation and accountability.

Tracing the trails of feedback data: How affected people pay the donors

We started this chapter with the story of Dolores, who never received a response to her text message. Her story motivated me to trace the journey of her message and the feedback databases more broadly. Dolores was by no means alone among our interlocutors in not receiving a response from the INGOs. What happened to all the feedback messages? Did anyone read them, and if so, what was the response? In the case of Dolores and several of our participants the feedback loop was not closed. Where did feedback data end up, if not with the people who originally offered their comments?

Back in the office of the main INGO that was based on the island of Sabay, our team met with some of the fourteen members of staff who worked in the accountability team. This is a very large team given the several urgent needs the INGO was dealing with. The accountability team were occupied with the following tasks: downloading messages into the database; coding messages according to the concern reported (e.g., sanitation, livelihood, cash assistance); flagging issues; and formatting the database. The head of the team was responsible for writing a report summarizing the issues and forwarding the report to the head of the local office, who in turn forwarded the data to the INGO office in Manila. From there, the datasets were forwarded to the INGO's headquarters in Europe. When I asked a local accountability officer whether they knew what happened to the data they spent so much time curating, they told me that 'once we send it to Manila, we never see [the data] again'.

Tracing the journey of feedback data, reveals that accountability data turn into audit trails that donors demand. The short cycle of funding means that humanitarian organizations need to apply for the renewal of funding every few months. Feedback data provide easy access to metrics to support funding applications. Some interlocutors pointed out that the quick turnover of staff further compounded the pressures on teams to use 'whatever data they could find'. Digitized feedback methods accentuate the tendency for 'accountability to affected people' to be turned into audit for donors. Digital platforms enable the easy aggregation of data into databases which, in turn, can be easily forwarded and reproduced. This would not have been as easy if feedback was in analogue form.

The extractive nature of feedback was acknowledged by several of my interlocutors from the 'digital humanitarianism project' who have worked in emergencies around the world. One interlocutor referred to feedback platforms as 'sucking platforms [as they] take from communities. But it should be feedback, not feed-from.' Another participant from the sector referred to 'zombie forms online' and added that 'feedback is extractive. [...] We package everything in a measurable way in order to serve our own agendas, and ultimately to appear better, more inclusive.'

The last quote reveals that feedback ultimately benefits humanitarian organizations, which are able to legitimate their presence and get funding to continue working. This is why accountability to affected people programmes have an almost ontological significance for the sector. As we saw in Chapter 1, accountability is a key metric to prove the impact as well as the impartiality of aid organizations – both of which are necessary for their funding.

It is not just relief organizations that benefit. Donors require robust audit trails to justify the international aid budget to their taxpayers. Finally, private companies benefit to the extent that their innovations get to be used 'for good' causes in humanitarian settings. This is the case with chatbots and 'AI for social good' projects, which often attract media attention, as we'll further explore in Chapter 4. The hype generated by novel technology applications in humanitarian settings provides publicity and branding opportunities for technology companies. Underpinning the use of commercial platforms such as WhatsApp or Facebook Messenger is the business model of social media, which extracts users' data for profit. Extraction takes place across many levels and for the benefit of different stakeholders, but the direction is always one way: from the majority to the minority world. Rather than heralding the participation revolution, accountability data confirm the geometry of power in humanitarian operations.

The harm of not closing the feedback loop

Audit has exploded in the neoliberal world. Schools, universities, hospitals, services, hotels, restaurants and bars are constantly ranked and assessed. Isn't it reasonable to expect that humanitarian operations will also be subject to audit? What is the harm if feedback data are used for the reporting to donors and for the renewal of funding?

The problem with accountability data turning into audit trails emerges when the feedback loop is not closed. Let's return to the example of Dolores, with whom we began this chapter. Dolores had to overcome the cultural barriers regarding accountability and 'aid as gift' shaped through the history of colonial encounters discussed earlier in the chapter. Dolores had to also overcome the difficulty of summarizing her complex predicament into 160 characters. And yet, despite surmounting these hurdles, Dolores never received a response. It was not surprising to hear that Dolores felt disaffected following her experience. We could see her disappointment when she realized that we were not representing the INGO she contacted. Later, Dolores told us that she would never use a feedback channel again.

Dolores wrote to the INGO to ask that her family is included in the livelihood assistance which so many of her neighbours received. She effectively complained about her exclusion from the aid distribution, which was a very common feedback theme. Earlier in the chapter we observed that such exclusions reflect local hierarchies and politics. In the case of Lia, her exclusion was the result of patriarchal values, which are steeped in the period of Spanish colonialism and the spread of Catholicism in the Philippines. Accountability to affected people initiatives were introduced to remedy the deficiencies of humanitarianism. By not closing the feedback loop, agencies do not fulfil the purpose of accountability and reform. But more worryingly, they also potentially harm people by making them feel disenfranchised. The silence of the humanitarian organization gave a clear message to Dolores that her problem was not significant enough for the NGO to respond to. Dolores and Lia felt that the agencies were part of a system that did not respect them. Not being included in the distributions was devastating. Not being listened to and not being given a chance to make your case added insult to injury.

When the logic of audit trumps the logic of accountability, it is not simply a matter of repurposing data for another objective. When the priority is the donors and the legitimation of the humanitarian projects, then the corollary is that people feel neglected. As one interlocutor from the aid sector put it, 'we instrumentalize people in order to appear better as an organization'.

Conclusion

Tracing the trails of feedback data across the humanitarian infrastructures has revealed the geometries of power in the humanitarian field. Despite purporting to constitute 'accountability to affected people', feedback data serve the logic of audit and marketization. The ultimate recipients of feedback data are the donors, typically Western governments, who demand evidence of impact in order to justify the budget of humanitarian assistance to their voters. The short cycle of aid projects exacerbates this trend. Humanitarian officers have to apply for funding, or the renewal of funding, every few months or even weeks in order to continue their operations.

Digital technologies and infrastructures are critical for turning accountability into audit data. The digitized nature of feedback and the replicability and reusability of data mean that messages can be easily aggregated into databases that can be seamlessly shared with donors and officers further up in the hierarchy of humanitarian organizations. The sharing is facilitated by existing digital infrastructures such as email and cloud computing. In other words, people affected by crisis pay for aid with their data. Digital feedback emerges as an extractive process, flowing from the crisis-hit areas in the global South to the global North.

The way that accountability is narrowly defined as feedback renders it a hollow process. Accountability is a very Eurocentric notion. In the Philippines there is no word for accountability, while local communities had not prior experience of being asked to give feedback. In the Haiyan fieldwork several of my interlocutors didn't fully understand the purpose of the feedback hotlines. Although feedback was meant to empower local communities and increase their participation, it became a form of epistemic violence by imposing a notion that did not resonate with the cultural context. It is indicative that several of the messages recorded on databases were 'thank you' notes. Coupled with the 'thank you' shrines dedicated to relief organizations in the wake of Typhoon Haiyan, these practices reveal that aid is filtered through colonial relationships of obligation and gratitude. The latest iteration of the technologized feedback comes in the form of chatbots, which select responses from a list of pre-scripted answers. By imposing a fixed menu of predetermined answers as feedback and by 'disciplining' people to speak in a chatbot-legible way, AI-programmes accentuate epistemic violence and coloniality.

The digitalization of feedback introduces exclusions (for example, those who cannot write or do not have good connectivity) and distancing. Several of our interlocutors only received automated messages that their message had been received, while some didn't even get those. Being treated with indifference by the machine, was experienced as demoralizing by those who overcame cultural and practical obstacles to participate in the feedback process. The predetermined answers imposed by chatbots suggest that feedback is now written even before people are invited to contribute their experiences. Chatbots crystallize what has always been the case with the narrow interpretation of accountability as feedback. Digital feedback systems have always tried to force messy and complex categories into small boxes. In so doing, 'accountability to affected people' initiatives offer only a veneer of participation. Even when they fail to bring about the 'participation revolution' promised by the 'Grand Bargain', feedback mechanisms succeed in achieving something important: they legitimate humanitarian projects.

The extractive nature of digitized feedback with its South–North trajectory and the way it imposes Eurocentric systems of knowledge with associated harms exemplifies technocolonialism. The parallel occlusion of power asymmetries under the pretence of participation is another distinctive feature of technocolonial practices.

In the next chapter we will observe that what is being extracted here is not just data, but also the value from experimentation with untested technologies in vulnerable settings.

4

Surreptitious Experimentation

Enchantment, Coloniality and Control

In March 2016, the Silicon Valley startup X2AI launched 'Karim', a psychotherapy chatbot, to support Syrian refugees in Lebanon. The chatbot uses natural language processing, a form of artificial intelligence (AI), to imitate human conversations in Arabic via existing platforms such as Facebook Messenger. Given the prevalence of mental health issues in situations of displacement and war (Charlson et al., 2019) and given the parallel lack of trained psychotherapists in such settings, chatbots have been proposed as a potential solution. This particular chatbot was reportedly piloted on sixty Syrians 'mostly men and boys', a small pilot for scaling up to a large population made vulnerable by conditions of protracted displacement: there are over one million Syrian refugees in Lebanon. X2AI developed the pilot in partnership with the 'Field Innovation Team', a non-profit specializing in technology in disaster-recovery, and the so-called 'Singularity University', the Silicon Valley business incubator and consultancy service.[1] 'Karim' is a pared-down version of 'Tess', X2AI's behavioural coaching chatbot available to people living in the US, which addresses depression and anxiety through existing instant messaging apps. While 'Tess' serves as a therapeutic tool that supplements, rather than replaces the role of a therapist, the roll out of 'Karim' wasn't accompanied by a parallel availability of professional psychotherapy.[2]

Karim is typical of the hundreds of technological experiments or pilots that have been taking place in the global South over the last couple of decades. As one of my interlocutors from the humanitarian sector put it, we are witnessing a true 'pilotitis'. The ubiquitousness of technological 'pilots' or 'experiments', several of which fall under the 'AI for good' phenomenon, is driven by the logics of solutionism and capitalism, but also those of securitization and humanitarian reform. Private entrepreneurs and digital developers are involved, but so are humanitarian organizations through their innovation labs. In Chapter

3 we observed that value is extracted from the data of affected communities for the benefit of humanitarian agencies and other stakeholders. In this chapter, I argue that value is extracted from experimentation with untested technologies in humanitarian settings for the benefit of private companies as well as NGOs and intergovernmental organizations. Contemporary experimentation in vulnerable settings in the majority world reworks the genealogies of medical and pharmacological experiments of the nineteenth and twentieth centuries (Petryna, 2009; Tilley, 2011). Just like historical experiments, contemporary 'AI for good' applications map onto unequal geometries of power and impose Eurocentric forms of knowledge in majority world settings. This is evident when large language models or machine learning algorithms are trained in the English language, but are implemented in non-English speaking settings.

This chapter observes that power asymmetries are further reproduced through the enchantment with technological innovation and experimentation. Because experiments build on technological infrastructures such as biometrics, which underpin humanitarian operations, we observe that experimentation is increasingly surreptitious. This means that experimentation is not readily acknowledged as such. The 106,000 people who initially participated in the *Building Blocks* pilot explored in Chapter 2, were not aware that they were taking part in an experiment. Because technological infrastructures are integrated into people's everyday lives, experimentation becomes diffused and blends into the background – a constant, yet inconspicuous experiment. Out of sight, yet in plain sight. All experiments rework the relations between those who conduct experiments and their subjects; surreptitious experiments, because of the unequal power geometries involved, have far reaching consequences that I explore in the final section of the chapter.

The chapter will focus on humanitarian chatbots as an exemplar of experimentation and technocolonialism. Chatbots were originally introduced in humanitarian settings in order to improve information dissemination and communication with affected communities. The Covid-19 pandemic accelerated the implementation of chatbots, as the need to deliver aid remotely increased due to social distancing measures. Since 2020, there has been a diversification of the uses of chatbots from education and skills-training, to psychotherapy and counselling. Chatbots are produced by non-profit and for-profit organizations bringing together several of the logics driving digital developments in the humanitarian space. Chatbots are also closely associated with the phenomenon of 'AI for social good'. For all these reasons the rest of the chapter will mostly refer to humanitarian chatbots – although I will also take the opportunity to return to earlier examples such as *Building Blocks*, which is one of the largest pilots in the humanitarian space having reached hundreds of thousands of refugees.

The chapter will proceed as follows. I will begin with a historical approach to experiments, which help explain why the global South is treated as a ready site for contemporary experimentation. I will then turn my attention to chatbots as an exemplar of experimentation in humanitarian settings. The remaining chapter, drawing on the 'digital humanitarianism project', will then address four major characteristics of technological experimentation in humanitarian settings: extractivism, coloniality, enchantment and the surreptitious nature of experiments.

Making sense of experiments

Experiments have been considered as a key scientific method (Popper, 1959). Just like science, experiments are synonymous with modernity, part of humanity's trajectory of inexorable progress. As scientific instruments, experiments are assumed to be objective and rigorous. Experiments conjure up images of laboratories and scientists clad in white coats working under controlled conditions. The laboratory is almost synonymous with experiments and its symbolism confers legitimacy to products. If something is 'lab tested' it is considered to have some kind of quality assurance. Traditionally, experiments were understood to be confined to laboratories and much of the literature on experimentation focuses on the laboratory as the research fieldsite (Latour and Woolgar, 1986).

Yet, a historical perspective reveals that experimentation has often extended beyond the science laboratory. In the nineteenth and twentieth centuries medical experiments took place in colonies where new drugs were tested before being rolled out in Europe (Latour, 1988). For Helen Tilley, colonial East Africa was a 'living laboratory' (2011) as the British colonizers via the Africa Research Survey applied 'modern knowledge to African problems'. For Achille Mbembe 'Africa is essentially [...] an object of experimentation' (Mbembe, 2001: 2). Adriana Petryna (2009) has documented the devastating consequences of pharmacological experiments for poorer countries where there is assumed to be a surplus of trial subjects. More recently, Noortje Marres and David Stark (2020) have observed that the transformation of testing from controlled laboratory experiments to ubiquitous testing in social environments has been facilitated by contemporary developments in computation and the availability of vast datasets. While I largely agree with their argument, I argue that experiments have taken place outside the laboratory well before computation.

It is already apparent that the claim to scientific objectivity typically associated with experiments is deeply contested. One of the chapter's main premises is that experiments are not neutral. Behind the veneer of objective science, experiments are steeped in ideologies, politics and vested interests. The field of tropical medicine was a tool of empire-building. Clinical trials in colonies, or former colonies, were largely conducted for the benefit of European people and for the advancement of Western medicine and pharmacology. Experiments are always more than just tests. They reconfigure relationships between scientists and people, and place human participants in positions of inequity. In so doing, experiments also produce subjectivities.

Experimentation in a historical context

The historical account of experimentation is necessary because it explains why the global South is readily available as a site of experimentation. In this chapter I take issue not with scientific experiments in general, but with those experiments that take place in relations of inequity.

To understand experimentation in the majority world, it is important to make sense of the relationship between science, empire and colonialism. Science was always central to the 'civilizing mission' of colonialism. Science and medical knowledge helped justify colonial expansion in the name of progress and enlightenment (Fanon, 1959). Europeans interpreted the infectious disease mortality in the colonies as 'a providential sign of the righteousness of the European imperial project and evidence of the superiority of white bodies' (Greene et al., 2013: 36).

Apart from serving as a moral justification, science was also a tool of empire. In Chapter 2 we saw how the emerging science of biometrics was used to control colonial subjects. There are several other examples such as the development of colonial medicine, which was key for facilitating the nineteenth-century 'scramble for Africa' (Farmer et al., 2013). Medical knowledge allowed Europeans to colonize almost all tropical Africa. Tropical medicine was developed to tackle diseases such as malaria and yellow fever in order to reduce the death rate of European officials in the colonies (Hochschild, 2019: 90). Sir Patrick Manson, the founder of the London School of Tropical Medicine, stated during the graduation of the first cohort of students: 'I now firmly believe in the possibility of tropical colonization by the white races' (cited in Greene et al., 2013: 42–3).

It is evident that science and medicine were more than tools; they were constitutive of colonialism in the sense that they helped 'conceptualize and bring into being the colonial project itself' (Suman, 2009: 375). Science and colonial medicine moulded subjectivities and entrenched the subjugation of the colonized. The disparities in disease mortality quickly translated into racial hierarchies based on inherent biological characteristics (Greene et al., 2013: 36). The establishment of tropical medicine as a distinct specialization of medical research and practice shaped popular stereotypes about the 'tropics' as a dangerous and unhygienic place and codified notions of racial difference (Greene et al., 2013: 41). For example, the US occupiers blamed the 1902 cholera epidemic on the Philippine people's unhygienic behaviours (Anderson, 2006). Anderson's research (2006) reveals that the oppressive public health measures during the US military occupation of the Philippines created the categories through which the American colonizers viewed Filipino people. Colonial medicine and racism have entwined histories. The dehumanizing nature of experimentation encapsulates the violence of colonialism (Mbembe, 2001).

This historical account explains why colonies, or former colonies, were assumed to be available spaces for the outsourcing of clinical trials. The history of the Pasteur Institute in France is a case in point (Latour, 1988). Via its network of branches in the majority world, the Pasteur Institute conducted trials in order to test drugs and vaccines. One such case was the tuberculosis BCG vaccine which was trialled in Algeria from 1930 and for a period of 26 years (Rosenberg, 2012). The Pasteur scientists designed a study to be followed 'like a laboratory experiment' (Rosenberg, 2012: 689). This involved 40,000 newborns in the poor and densely populated Algiers Kasbah, 20,000 of whom would be vaccinated (and then revaccinated at specified intervals), with the rest being the control group. The BCG Algiers trial took advantage of the subjects' poverty: no informed consent was sought, personal records 'were violated', while researchers saw the study as a 'service to the local population' with no regard to safeguarding such a vulnerable group (Rosenberg, 2012: 692). In colonial settings the boundary between experimentation and care has always been fuzzy (Anderson, 2006).

The Algiers BCG trial was driven by desire to amass empirical evidence to shore up the new tuberculosis vaccine which had been criticized for not having been founded on robust statistical evidence regarding its efficacy. The experiment aimed to provide data to satisfy the statistical requirements set by the League of Nations. By moving the trial to Algiers, the scientists were able to bypass the practical considerations and administrative hurdles associated with a trial in France (Rosenberg, 2012). The Algiers trial exemplifies the notion of colonies as 'laboratories of modernity' (Rabinow, 1995).

If the Algiers trial benefited the reputation of the Pasteur Institute and its scientists, more recent experiments, serve the logic of capitalism. Following the expansion of big pharmaceutical companies after the Second World War, there has been an increased demand for clinical trials. Petryna's study of the clinical trials industry documents their impact among subjects in middle- and low-income countries (Petryna, 2009). A disturbing example of pharmacological experimentation was the Trovan scandal in Nigeria, which trialled an untested antibiotic among children without due protocol, informed consent, nor ethics (Petryna, 2009). A local meningitis epidemic in 1996 gave the pharmaceutical company Pfizer access to subjects, while the public health emergency was used to justify the very questionable arrangements (Petryna, 2009: 40). Given the complete lack of heath care, the clinical trial was welcomed as an opportunity to provide care – even though the drug was untested. When five children died, the lawyers acting for Pfizer argued that the company was providing care – rather than conducting an unethical experiment (Petryna, 2009: 39). We notice again the same slippage between the language of care and experimentation that was used in the Algiers BCG trial and in earlier colonial experiments. Although there have been thousands of clinical trials among European and North American populations (often among disadvantaged communities) it is hard to imagine a contemporary study in the minority world context without clear protocols and a regulatory framework.

The colonial genealogy of science and experimentation informs contemporary practices around technological pilots. It explains why contemporary experiments and technological pilots take place among marginalized people in the global South. From the discussion so far, we can discern two further themes that are relevant for our analysis. The first is that experimentation is a process of extraction. Apart from direct profits for commercial companies, experiments produce value that benefit global North organizations, as happened with the Pasteur Institute. In the case of technological experiments, the list includes technology companies, entrepreneurs, donors, governments and, of course, humanitarian organizations – all based in the minority world. Second, experiments in asymmetrical settings such as refugee camps reproduce the structure of their colonial counterparts. I argue that technological experiments such as chatbots, part of the 'AI for good' trend, reproduce the coloniality of power (Quijano, 2000). In order to make sense of chatbots, we first need to consider AI.

Artificial intelligence

AI refers to several phenomena including computational approaches (for example, machine learning, which in turn can include large language models, artificial neural networks, or natural language processing); and practices, such as design or coding. AI is also a 'planetary infrastructure' (Crawford, 2021) from the extraction of resources and the submarine cable infrastructure, to cloud computing and the e-waste of defunct devices. Crucially, AI is also a form of knowledge production and a social imaginary (Taylor, 2002) as evidenced by the utopian and dystopian narratives about its capabilities. Narratives about AI influence the design of new applications (Natale, 2019), as well as their public perception and regulation. Given the opacity around the workings of algorithms and automation (Pasquale, 2015) the public's understanding of AI is inevitably shaped by popular culture narratives (Craig et al., 2018). The term 'artificial intelligence' is loaded with meaning, implying a 'thinking' machine, even in cases where there is little evidence of autonomous thinking. Anthropomorphism, the attributing of human characteristics such as intelligence to software programmes (for example, when asserting that 'the machine *learns*') carries powerful connotations of agency. The phrase 'the machine thinks', or 'the machine learns' performs an important task in framing technology as socially agentic. At the same time, the words 'artificial' and 'nonhuman' imply a distancing from humans that occludes the human labour behind computation (Anwar and Graham, 2020a) and cloaks technologies in an aura of objectivity.

Part of the dominant AI imaginary is the belief that as 'intelligent' and scientific, it is more advanced than previous forms of computation or innovation. This is typical of a teleological understanding of technology, where the latest iteration is claimed to be better and more advanced than the previous ones (Kember and Zylinska, 2012). As we have already seen in the Introduction, the 'AI for social good' phenomenon exemplifies the mythology of 'artificial intelligence' as inherently progressive. The development of humanitarian chatbots illustrate some of these issues.

Chatbots

Chatbots have captured the public imagination since the arrival of ChatGPT, a large language model-based chatbot developed by OpenAI, in November 2022.[3] The response to ChatGPT has triggered some of the most optimistic and most dystopian views on artificial intelligence. In any case, the hype around ChatGPT and generative AI display an enormous enchantment with technology. While ChatGPT and other large language model chatbots dominated the headlines in 2023, most chatbots don't have the capabilities of large language models.

Chatbots are essentially software programmes that can recognize text, or voice-based inputs and interact with humans online (Gehl and Bakardjieva, 2017). Chatbots have a long history going back to ELIZA, the first chatbot created by Weizenbaum (1966) at the Massachusetts Institute of Technology in the 1960s. The idea of the 'thinking machine' can be traced back to Alan Turing, who in 1950 argued that computers were capable of 'intelligent behaviour'. The 'Turing test', originally known as the 'Imitation Game', aims to prove whether 'machines can think' in the context of conversations (Turing, 1950). To pass the test, a machine has to produce responses that are indistinguishable from those of a human.

The degree to which AI applications can produce human-like speech varies dramatically (Broussard, 2018). Low-level chatbots, which are still very prevalent globally, provide answers from a list of predetermined questions. Such chatbots can be useful in handling 'frequently asked questions' where questions and answers are pre-scripted, but they can't respond to more complex questions nor hold a conversation (Bakardjieva, 2015). Advanced chatbots on the other hand, run on natural language processing and machine learning algorithms which aim to analyse and imitate human language. The latest iteration of advanced chatbots, exemplified by ChatGPT uses large language models to generate text. Most humanitarian chatbots are low or mid-level chatbots. While some will only handle 'frequently asked questions', others will respond to prompts. Chatbots have been developed by humanitarian organizations, such as UNHCR, the WFP and IRC, as well as by private companies, such as X2AI, which launched the psychotherapy chatbot 'Karim'.

Extractivism

Technological experimentation in the humanitarian sector is driven by most of the logics we have already encountered in previous chapters. For example, chatbots have been introduced in order to improve communication with communities (logic of participation and accountability), but they are also driven by the push for efficiency savings by humanitarian organizations. Technology companies are key players here as most chatbots run on commercial platforms such as WhatsApp. Technology companies are also involved in chatbot design as in the case of Karim. Experimentation offers significant branding opportunities to technology and other commercial companies. Innovations such as chatbots can generate hype and publicity around particular products, which ultimately benefits the parent company. Pilots can be opportunities for companies to test products before rolling them out commercially. Crucially, the logic of solutionism underpins the 'pilotatis' mentioned by one of my interlocutors. There is a whole industry of experimentation that encompasses for-profit and non-profit companies, hackers and entrepreneurs as well governments. Most humanitarian chatbots have been developed in hackathons. In the following paragraphs I analyse technological pilots as a process of value extraction by humanitarian organizations and by technology companies.

Technological pilots are often conceptualized in hackathons or 'bootcamps' in Europe or North America. These are attended by entrepreneurs, private companies, coders and representatives from humanitarian companies. Between 2015 and 2018 – during the so-called European refugee crisis – over 1,500 apps for refugees were launched (Leurs and Smets, 2018) many of which were designed in the thousands of hackathons that took place in Europe and beyond. Prominent institutions such as the Vatican organized the VHacks hackathon in 2018 in the Vatican City. Techfugees (officially the Techfugees Foundation) is a for-profit private company[4] established in 2015 in the UK to develop technology solutions for the inclusion of displaced people.

The larger technological pilots are the result of private–public partnerships between humanitarian organizations and commercial companies. Many of these partnerships are forged in events organized by the innovation labs of agencies or INGOs. For example, the WFP's Innovation Accelerator, which has been at the forefront of such developments, holds a 'bootcamp for start-ups and social entrepreneurs, who believe they have what it takes to help end hunger'.[5] The bootcamp is the first point of contact for start-ups and the WFP. After the bootcamp, successful companies receive 'funding, hands-on support and access to a global network to bring their concept to life'.[6] A WFP representative described the process during an industry event as follows:

> After the bootcamp the best start-ups can get funding from us from 50,000 to US $100,000 to pilot the idea in a developing world context. Our role is to do the matching . . . we will connect you to one of our offices around the world and part of the funding will be to go and test it on the ground in that field capacity ... We will continue to give you mentoring and help you think of your business model as you go forward.[7]

This quote exemplifies the observation in Chapter 1 about solutions seeking problems. Bootcamps exemplify the solutionist logic that begins with an innovation or an idea and then aims to apply this solution to a region in the world. The region is 'matched' to the solution, rather than the other way around. Instead of starting with the issues a community faces and then identifying potential approaches – including ones grounded on the locale – bootcamps begin with 'concepts' which are 'tested' on local communities with a view to 'scaling up' through a clear 'business plan'.

I will address this top-down, Eurocentric approach in the following section. But what is clear here is that humanitarian settings are seen as laboratories for trialling untested technologies, often sanctioned by the organizations that operate in care of these populations. It is important to distinguish between different kinds of pilots here. While some pilots only involve the trialling of new methods of farming (for example, hydroponic systems that allow vegetables to grow in arid areas),[8] other pilots involve the participation of humans. It is these latter projects that concern me here.

By trialling untested technologies such as chatbots, private companies extract value to improve the design of their products. Pilots are also selected on the basis of a viable business plan. At the same time, pilots help companies generate publicity and hype around their products. As one of my interlocutors from the aid sector remarked:

> We have this interesting situation that we're in […] when companies invest a lot of time and effort for products that they haven't yet found a market for. […] And there's a lot of effort trying to create hype. The humanitarian sector is one of the ways of creating headlines. This is essentially advertising for companies.

Technological pilots provide excellent branding opportunities for companies. The example of the Karim chatbot with which we opened the chapter provides an apt example. The pilot featured on a range of media outlets, from the *New Yorker* to the *Guardian*[9] providing significant publicity for the company X2AI (recently renamed as Cass.AI), which offers 'behavioural health coaching' via 'AI mental health assistants' in the US. In the rapidly expanding 'AI for good' sector, it is crucial to be seen to have products that suggest a 'caring brand'.

Branding and visibility are not just relevant for commercial companies. In an increasingly marketized environment, where humanitarian organizations compete for funding, humanitarian organizations aim to build a distinctive identity to attract donors. As one of my interlocutors put it when discussing an app that their team had developed: 'Part of our motivation was to improve our visibility – to show to other organizations what [our agency] does.' During my fieldwork, I often encountered the view that humanitarian organizations were bureaucratic and slow. This was usually expressed most strongly by the private companies and entrepreneurs involved in humanitarian projects. Technological innovation, because of its association with progress and the future, was seen to offer an opportunity to replace that traditional image with a 'dynamic and relevant' one, which is attractive to donors.

Competition for visibility, also takes place within large organizations, where different divisions compete for funding from the centre. Speaking about another app, one of my participants remarked:

> Our main purpose through these projects, I'm not going to lie, was to prove a point [within the organization] or to skill up globally, nationally or regionally. [The app] was an example of proving a point.

The above motivations and justifications for projects explain why several pilots are very short lived. If the aim is to 'prove a point' or to attract visibility, then the brevity of projects should come as no surprise. Another reason that explains the low sustainability of technological pilots is their weak grounding in local communities, which is unsurprising, given the top-down direction of innovation. Of course, some pilots such as *Building Blocks*, one of the largest and most successful initiatives of the WFP's Innovation Accelerator, have reached hundreds of thousands of people – although the success of that programme is explained by how well it satisfied several of the logics of digital humanitarianism. The short span of many of the pilots does not mean that they are less extractive. Ironically, the shorter the life of a pilot involving human subjects the more extractive it is likely to be – as the short span may suggest a short-term goal (branding, visibility, 'proving a point'), which is often unconnected to the issues local communities face.

Even the shortest pilots involve a further kind of extraction if they use social media platforms. Most chatbots run on popular messaging applications such as WhatsApp and Facebook Messenger, both owned by Meta. Many of my interlocutors argued that it makes sense for aid agencies to use the platforms that people already use. This is a compelling point, but it ignores that by sanctioning platforms that have questionable data safeguarding practices, humanitarian agencies inherit the concerns surrounding the business model of social media companies, whereby users' data are extracted for profit (Zuboff, 2019). By using WhatsApp and other messaging apps, the safeguarding and privacy of the data of humanitarian subjects is outsourced to Silicon Valley companies. As Sean McDonald argues, by relying on Facebook or other messaging apps for essential services, humanitarian agencies extend their remit ('the imperative to do good' and 'do no harm') to private technology companies (McDonald, 2019). If a refugee relies on WhatsApp for their own private use, this is very different from only being able to access vital information regarding food distributions via a WhatsApp chatbot (Madianou, 2021). Because most humanitarian pilots use the WhatsApp API without formal agreements with Meta, they have no control over the safeguarding of the data and metadata, which are equally sensitive when dealing with persons of concern (ICRC and Privacy International, 2018). Even in cases when humanitarian organizations sign formal partnerships with big technology companies it is not clear 'what leverage they have to ensure the enforcement of that control' (McDonald, 2019: 4).

Technological experiments involve the extraction of people's data, as well as the extraction of further value from knowledge production (for example, the insights from the testing of the product), the labour of the refugees who test the products, the visibility bestowed on companies and organizations and the branding opportunities. These are not mutually exclusive. Extraction can occur at different levels – data harvesting can exist in parallel with the extraction of visibility, or labour. Multiple stakeholders can benefit simultaneously. A start up can extract value from the labour of refugees, a humanitarian organization can extract visibility, while a social media company can extract data. An infrastructural perspective captures the flow of data and value across multiple stakeholders. Further, the distinctions between humanitarian organizations and technology companies are increasingly blurred. We have already seen that technology companies develop their own pilots – which turns them into de facto humanitarian organizations, but without the requirements to adhere to the humanitarian principles of humanity or neutrality.

Conversely, when humanitarian agencies, through their innovation labs or accelerators, support the development of a chatbot, then they become technology companies. Most pilots involve private–public partnerships. *Building Blocks*, one of the largest WFP pilots ever, is run by a private company, IrisGuard, which specializes in biometric solutions for border security and humanitarian aid[10] raising critical questions about function creep. The complex supply chains of technological experiments blur the lines of responsibility and accountability. Extraction is never just a matter of profit or accumulation. It involves clear risks and harms that map onto the coloniality of power (Quijano, 2000).

The coloniality of technological experiments

The geographic distribution of technological pilots reveals the global geometries of power. A survey of over one thousand 'AI for social good' applications between 2008 and 2020 found that the vast majority of projects were US-based, while the few that focused on Africa were directed by US institutions (Shi et al., 2000: 45). My survey of the forty-nine AI applications in the humanitarian sector in 2021 revealed that they were all led by organizations in the minority world (predominantly North America and Europe).[11] Hackathons, which are key sites for the design of new applications, typically take place in cities in the global North. Most chatbots are designed in hackathons that are far removed from the reality of the humanitarian settings. The World Food Programme, which has been at the forefront of chatbot development, developed its Foodbot in a hackathon in January 2017 in New York City – organized by the global marketing and data analytics company Neilsen on behalf of the WFP. The hackathon was attended by 'developers, students, volunteer hackers and Neilsen staff' (DIAL, 2018: 9) – presumably all residents of New York or other global cities. This reflects a point upon which many of my interlocutors from the aid sector agreed: hackathons are mostly white and male with little or no refugee participation. During my fieldwork between 2016 and 2021, I attended over ten hackathons, most of which had no refugee participants. One of my interlocutors, active in the 'technology for good' space, asserted that he couldn't 'remember one hackathon [...] that actually had refugees in it'. Another participant echoed this observation in the following quote:

> There was this big hackathon for refugees in a European capital [...] there was like 200 people from Latin America and Europe, no refugees, even no refugee representatives. It was mostly journalists and designers and programmers, who were tasked for five days to invent solutions to the refugee crisis. And I'm like that's not at all how it works. Obviously, the project that won was an artistic installation [...]. People did what they knew. [...] Another project, which didn't win, was a 'marry a refugee app'. Like if you're from Germany, so you can maybe give your support to a refugee by marrying them. [...] It was so horribly tone deaf.

While I observed a slight shift from late 2019 towards holding a small number of hackathons in the global South, the vast majority were still held in the minority world. While there is increasing awareness about the need to include refugees and other affected people in hackathons in the global North context, this often feels tokenistic. Technological pilots may have proliferated, but the direction of travel remains one way: from the minority to the majority world. This is an example of 'innovation without participation' (Scott-Smith, 2016). The fact that most innovations are conceptualized and developed in the global North before being implemented in humanitarian settings, which are typically in the global South, has important implications.

The geographic unevenness of technological pilots matters because it reveals the power dynamics involved. Computation depends on classifications, which, as we saw in previous chapters, are inherently political and often reflect the dominant values of the environment in which they were created (Gebru, 2020). Because classifications are embedded in infrastructures, they become invisible, which renders them even more powerful (Bowker and Star, 1999). In her ethnography of technological innovation for development in India, Lilly Irani argues that 'design asserts categories with racialized and colonial genealogies' (2019: 218). Silvia Lindtner's ethnography of technological innovation in Shenzhen observes that 'technological experiments in China were driven by [...] the colonial othering of Chinese people that both their government and the West upheld', reminding us yet again how coloniality is often internalized by local elites (Lindtner, 2021: 48). 'Design designs', as Arturo Escobar has put it (Escobar, 2018: 4). In other words, because it reflects ways of seeing the world, design 'creates ways of being'.

Such practices explain why automation and AI systems reproduce and heighten inequality (Eubanks, 2017) and racial and gendered discrimination (Benjamin, 2019; Noble, 2018).

Given that the design and management of 'AI for social good' projects is located in the global North it is not surprising that the values informing some of these platforms are Western-centric. The language in which chatbots operate is vital here. Decolonial writers like Ngũgĩ wa Thiong'o have stressed how 'language carries culture and culture carries [...] the entire body of values by which we perceive ourselves and our place in the world' (1986: 16). Language is one of the most fundamental tools that reproduces power asymmetries. With few exceptions, most humanitarian chatbots operate in local languages.[12] Still, even when a chatbot operates in Arabic, the machine learning algorithms have been largely trained in English language datasets. In other words, when a chatbot operates in Arabic, this often entails a translation from English. The infrastructure of the chatbot (e.g., the artificial neural network) is based on the English language and the classifications are decided by the designers – possibly English-speaking, although not exclusively – in the minority world. The question of language is particularly relevant in the case of medical chatbots which depend on classifications regarding health and illness which are culturally specific.

The importance of cultural sensitivity is even more vital for mental health chatbots. Awareness of cultural norms and idioms is crucial in psychotherapy especially as sociocultural factors contribute to mental health and illness. Given the cultural specificity of emotions, a deep understanding of cultural codes and norms is necessary for psychotherapy to be successful. The same applies to gender, sexuality and social class which, too, are culturally specific and intersect with mental health. Cultural sensitivity cannot be achieved by translating code into different languages. What matters is understanding how emotions like shame are expressed in a particular cultural context, something that necessitates an intimate understanding of the local culture. Karim, the mental health chatbot, was piloted in Lebanon – albeit among a very small sample – but the application, was conceptualized and programmed in California as part of the portfolio of the company's psychotherapy chatbots. A further issue concerns the way trauma is conceptualized in this context. This is a broader critique of psychiatric classifications by postcolonial scholars, who have questioned the Eurocentric construct of post-traumatic stress disorder (PTSD) as centred on a single traumatic event (Fassin and Rechtman, 2009). The causes of trauma are structural, and therefore trauma is collective and, in the case of refugees, ongoing. There is no 'post-trauma' in situations of protracted displacement.

Integrating the chatbot with a network of locally trained psychotherapists, who can intervene when necessary, can improve the limited cultural sensitivity embedded in the code. Yet, 'Karim' was not accompanied by any actual therapists, unlike 'Tess', the AI company's behavioural coaching chatbot, which is available to US customers.[13] That a bare-bones version of the mental health app was considered suitable for a population living in conditions of extreme precarity speaks volumes about the colonial mindset of the design process. Imposing a Eurocentric understanding of mental health through code and design is a form of epistemic violence that dehistoricizes trauma and asserts the coloniality of power (Fanon, 1952; Quijano, 2000).

How ironic it is then when the promotion material for chatbots claims to 'decolonize humanitarian information'. This phrase is used in the promotion of a different chatbot, developed by the IRC and Mercy Corps in collaboration with the technology company Zendesk. Signpost is a chatbot that provides practical information to people impacted by emergencies via popular platforms such as WhatsApp. The first iteration of the chatbot, Refugees.info, was launched in Greece in 2015 when over one million refugees from Syria and elsewhere made the crossing to Europe. To be fair, because Refugees.info provides relatively straightforward information to simple questions such as where to find hospitals or asylum application offices, the question of cultural sensitivity is not as critical as it is with health or educational chatbots. But, to equate decolonization with the dissemination of practical information is a form of 'decolonial washing' – a branding exercise that depoliticizes decolonization. There is nothing here that is unsettling or which attempts to change the problematic social order of colonialism to justify the use of the term decolonial (Tuck and Yang, 2012).

People in the majority world have often been presented with the dilemma of having to choose between no service or a lesser, or perhaps less safe, service. Mark Zuckerberg encapsulated this flawed rationale in his infamous statement that 'having some internet connectivity is better than none' when trying to justify the problematic programme of Facebook Free Basics or, as it is called at the time of writing, Discover. Internet.org which is how Free Basics / Meta Connectivity was originally known, was marketed as access to the internet for the world's most disadvantaged populations. Instead, the platform only gave access to a pared-down version of Facebook plus a handful of selected apps. Internet.org or Free Basics or Meta connectivity (the different names through which the programme has been known) are contrary to the Internet's fundamental principle that it is open to all. What Free Basics conceals is that it is a strategy to increase the company's market share by expanding into the global South market. It is telling that Internet.org was initially branded as a 'humanitarian project' and promoted in a speech Mark Zuckerberg gave to the UN General Assembly.[14]

Another significant issue is the question of data safeguards especially given the heightened vulnerability of humanitarian settings. As we've already discussed, technological pilots run on existing platforms such as Facebook Messenger or WhatsApp thus contributing to the extraction associated with the business model of social media. Of greater concern is the potential risk of misinformation if, for example, an app is out of date and offers mistaken information. During my 'digital humanitarianism' fieldwork, I took every opportunity to interact with every open access chatbot available and was able to witness out of date information offered on several occasions (Madianou, 2021). If a refugee is given wrong dates or other out-of-date information the consequences can be severe. As one of the NGO interviewees put it, 'if a chatbot directs refugees to the wrong meeting point, and it has taken them hours and money to get there this can be very problematic'. Misinformation, however unintentional, can be at odds with the humanitarian imperative of 'do no harm'. This raises the issue of responsibility in 'human–machine communication' (Gunkel, 2018). Who is responsible if a chatbot disseminates out-of-date information? Who can be held accountable? This is why the same interviewee remarked: 'You've got to have a human there, who can step in. But given these apps are driven by efficiency concerns, I'm not sure they will be prepared to resource it'.

The enchantment of technology: Magic and power

George Williams, in his 1890 Open Letter to King Leopold II of Belgium, outlined the major charges against the most brutal colonial regime of its time. One of the main accusations concerned the use of science to dazzle the Congo chiefs and make them sign their land over to Leopold. Here's an extract from Williams' fiery Open Letter where he describes the use of tricks by the notorious explorer Henry Morton Stanley who signed the 'treaties' on behalf of King Leopold:

> A number of electric batteries had been purchased in London, and when attached to the arm under the coat, communicated with a band of ribbon which passed over the palm of the white brother's hand, and when he gave the black brother a cordial grasp of the hand the black brother was greatly surprised to find his white brother so strong, that he nearly knocked him off his feet ... When the native inquired about the disparity of strength between himself and his white brother he was told that the white man could pull up trees and perform the most prodigious feats of strength [...] By such means ... and a few boxes of gin, whole villages were signed away to your Majesty. (Cited in Hochschild, 2019: 109–10)

If experiments are predominantly associated with images of laboratories, a flip side is the image of magic. Parallel to the imaginaries of numbers and science, contemporary imaginaries of AI are equally shaped by the association of technology with magic and religion. From the advent of electricity (Marvin, 1990) and the telegraph (Supp-Montgomerie, 2021), technology has been associated with enchantment and spirituality. Rather than conceiving of science and magic as incompatible, the boundaries between science, magic and religion are more permeable than often assumed (Styers, 2004).

As we can see from the above extract, science and technology have historically been used to dazzle. While contemporary uses of technology and experimentation do not inflict the horrors and injustices associated with the brutal colonization of the Congo, casting technologies as enchanted objects continues to reflect and reproduce power asymmetries.

In 'The Technology of Enchantment and the Enchantment of Technology' Alfred Gell was interested in how artifacts 'cast a spell' over people by 'functioning as weapons in psychological warfare' (Gell, 1992: 44). The example he used was the intricate and imposing canoe prow board from the Trobriand islands, which is exquisitely crafted to dazzle the Trobrianders' overseas partners so that they would be persuaded to pay more for the Trobrianders' goods than they might otherwise be inclined to do. The dazzling prow board made the overseas trading partners 'take leave of their senses and offer more valuable shells or necklaces' to the Trobriander sea merchants (Gell, 1992: 44). With its intricate engraving and adornments, the prow board enchants and helps the Trobrianders achieve their goal of maximizing the value of their trade. The canoe-board confers magical prowess on the owners of the canoe, whose power and status are, in turn, heightened.

'The enchantment of technology' (the hold technologies have on us) depends on 'the technology of enchantment' – the actual making of artifacts (Gell, 1992). Artifacts need to be crafted – but their enchantment depends on the erasure of the labour behind their construction. Crafting entails erasure. Just like a poet needs to erase the hard work behind the crafting of verse, the maker of artifacts hides the laborious process of carving, sawing, planing, engraving and painting. The 'technology of enchantment' is not just about highly skilled, hard work; it is also about being able to make the final artifact appear effortless, almost natural. It is only by hiding the work through which it is produced, that the magical efficacy of technology is achieved (Gell, 1992).

These observations can be applied to media technologies. Technological mediation depends on the erasure of its own work. Eisenlohr highlighted the 'propensity of media to erase themselves in the act of mediation' (2011: 9). Media become so entangled with what they mediate that they become invisible (Meyer, 2013: 312). The propensity of mediation to erase its own work is referred to as 'immediacy' by Bolter and Grusin (2000). Erasure does not mean that technology is removed or obscured. In fact, it is the combination of technological affordances that creates the illusion of immediacy. Bolter and Grusin have argued that immediacy works in parallel with hypermediacy – the awareness of technological affordances. VR (virtual reality) exemplifies the convergence of immediacy and hypermediacy. The combination of affordances and technological propensities (hypermediacy) aim to create an immersive experience that offers the illusion of immediacy. VR depends on complex (and still expensive) hardware and software in order to create a user experience that appears to be devoid of technological mediation. The paradox here is that immediacy (the appearance of no mediation) depends on heightened hypermediacy (the application of complex, converged technologies).

The opacity of algorithms and software further exemplifies the erasure of mediation. Even though algorithms support most online activity, from a simple web search to how we view content on social media, they remain unknown and mysterious. Pasquale (2015) uses the 'black box' metaphor to refer to algorithmic systems whose working remains hidden. Because computation is concealed, but its outputs are presented as natural facts ('givens'), algorithms and automation are shrouded in an aura of enchantment. This is one of the reasons why Chun (2011: 87) describes software as 'magic'. Algorithms and automation are the perfect enchanting technologies.

The enchantment of technology does not imply that technologies are perfect. To return to the example discussed in Chapter 3, humanitarian chatbots often fail the test of intelligence. Rather than fully fledged conversational agents that can mimic natural language and respond to complex questions, humanitarian chatbots are programmed to provide scripted answers from a list of 'frequently asked questions'. And yet chatbots are suffused in the aura of enchantment. Their enchantment relates to the way that the technologies are fetishized as powerful – not the degree to which they fulfil their promises. Enchantment operates at three levels: Western publics; the humanitarian sector, including donors; and affected communities.

The case of 'Virtual Reality (VR) for good' captures how enchantment operates at all three levels. VR technology has been used as a way of engaging 'Western publics' with humanitarian campaigns, including fundraising. By immersing Western publics in the realities of refugees, VR is championed as the 'empathy machine' by its producers (Gruenewald and Witteborn, 2022; Ponzanesi, 2024). In 2015 the UN commissioned the VR film 'Clouds Over Sidra' to showcase life in the Za'atari camp through the eyes of a young girl, Sidra. The film received significant publicity in the mainstream media, and has been shown in several industry events as a way of attracting donor funding. Amanda Hope Macari describes how the film producers recalled how the Za'atari residents assembled to observe with awe the intricate VR camera and equipment (Macari, in preparation). The presence of awe-inspiring technology further accentuates the power imbalance of humanitarian filmmaking, with largely white, male crews filming racialized 'others' made vulnerable by displacement and dispossession. In this sense, VR filmmaking reworks the colonial genealogies of ethnographic filmmaking and increases the inequities between filmmakers and filmed subjects.

The enchantment of technology is also evident when pilots receive awards, as happened with the 'Refugee Text' chatbot, which was selected for inclusion in the '2018 Designs of the Year exhibition' at the Design Museum in London. The now defunct chatbot, which was produced in 2016 as a response to the informational needs of refugees arriving in Europe, was an example of innovation with a short shelf life. Yet, through its selection and display in a high-profile exhibition, the chatbot acquired enchanted status as an item of cultural and social significance. In 2023 UNHCR won the 'Best Impact Project Award' at the Paris Blockchain Week, the largest blockchain industry event, for a pilot project using blockchain technology to disburse cash to people displaced or impacted by the war in Ukraine. This pilot is similar to *Building Blocks* but uses stablecoins, a form of cryptocurrency, when transferring funds to the recipients' 'digital wallets'. The enchanted objects confer power to their owners: the humanitarian organizations or the private companies and their philanthropic departments. The 'enchantment of technology' also explains why 'AI for good' is increasingly appropriated as a marketing and branding strategy.

There are parallels with the notion of 'charisma', which Morgan Ames uses to analyse the 'uncanny holding power of technology' (2019: 190). Her ethnography focuses on the much-hyped 'one laptop per child' (OLPC) programme, which was launched by Nicholas Negreponte in 2005. Providing children in low-income countries with cheap laptops was hailed as a way of accelerating young people's education and lift them out of poverty. Ames details the hyperbolic technological promises of the OLPC programme, but also how these failed to translate into actual outcomes for children in Paraguay. Yet, the aura of charisma meant that criticizing the project was often treated as an act of heresy (Ames, 2019: 189–90). Just as technological enchantment reproduces power hierarchies, the OLPC charisma ended up 'reinforcing the very inequities it meant to alleviate' (Ames, 2019: 191). By mirroring the personal and professional identities of its creators, the charismatic technology blinded them to criticisms of racial and gender bias or other issues of social justice (Ames, 2019: 191). 'Blind spots' are a pattern in projects driven by technological solutionism and the faith in the 'disruptive' power of technology (Sims, 2017: 11).

Placing an 'enchanted technology' in a very asymmetrical context amplifies existing power imbalances. By conferring authority on humanitarian organizations and for-profit companies, chatbots rework and revitalize the existing asymmetries of humanitarianism and reproduce the coloniality of power. Humanitarian AI may not be an example of 'psychological warfare' (Gell, 1992), but it has the potential to create hierarchies and boundaries between the designers and intended recipients of technology. In the case of the Design Museum exhibition, the clear winners are the chatbot developers, who gained visibility and distinction through their inclusion in a high-profile exhibition. This contrasts with the invisibility of refugees, who remain stuck in camps in Greece and Italy and whose plight provided the raw material for the elevation of the chatbot designers. As it dazzles, the enchanted object silences any critical voices.

Surreptitious experimentation

One of the largest technological pilots is *Building Blocks*, the WFP biometrics and blockchain-based cash distribution system, which we have already encountered in Chapter 2 and elsewhere in the book.[15] *Building Blocks* started in 2017 as a pilot of 10,000 people before expanding to 106,000 refugees in Jordan. In 2021 it scaled up to over one million users in Bangladesh and Jordan. In Chapter 2 we observed the infrastructural character of *Building Blocks* as it draws on the vast biometric databases of UNHCR and the WFP in order to authenticate refugees and allow them to use their cash assistance in designated grocery stores.

I argue that a new type of experimentation is enabled by digital and computational infrastructures. Because infrastructures operate in the background, they are invisible and taken for granted. When experiments take advantage of digital infrastructures, they, too, become invisible. In these circumstances, experiments become diffused. When a pilot occurs in what is an everyday, vital infrastructure, then participants may not be aware that they are taking part in an experiment. This compounds the already problematic consent issues discussed in Chapter 2. For example, if an experiment mobilizes the foundational infrastructure of biometric databases and essential cash distribution systems, then how will participants know that a pilot is taking place and how can they even opt out? It is not possible to opt out of one's own foundational data and it is not possible to opt out of aid-in-cash assistance when there is no alternative source of income.

Biometrics and blockchain are not the only infrastructures that enable experimentation. Chatbots are also infrastructural insofar as they rely on messaging apps and mobile phone devices, which they almost invariably do. Mobile phone and social media infrastructures are also vital, indispensable and invisible thus providing a perfect space for technological pilots. I argue that we are witnessing a new type of experimentation, which I call 'surreptitious experimentation'. This involves a continuous flow of experimentation which is not named as such and which operates in the infrastructural background, and therefore remains hidden – and yet in plain sight. Infrastructural experimentation exemplifies Marres and Stark's (2020: 428) observation that the social environments are modified so as to enable experimental operations. Infrastructures allow for more experimentation to take place.

Experimentation facilitates the changing of the settings – and not just the testing of the settings (Marres and Stark, 2020). The consequence is that experimentation becomes a form of control. The analysis of the *Building Blocks* pilot in Chapter 2 revealed how the system ultimately controls the movement and the spending pattern of recipients. Similar observations about how cash assistance programmes effectively control the movement of refugees have been made by Martina Tazzioli who studied UNHCR prepaid debit cards programme in Greece (2019). In surreptitious experimentation the aim is no longer to learn and to observe, but to intervene into the environment and to control the experiment subjects (Marres and Stark, 2020).

Experiments have always reconfigured the relationships between those tested and those testing, or between 'problem owners' and 'problem solvers' to use the language from the UN's 'AI for good' Global Summit.[16] All experiments position subjects into unequal positions. Surreptitious experiments accentuate this inequity and further compound the asymmetries stemming from extraction, coloniality and enchantment.

Ironically, even though technological experiments extend the genealogy of clinical trials, they are far less regulated than the latter. Clinical trials are now tightly regulated by bodies such as the Food and Drug Administration (FDA) in the USA, which approve pilots after rigorous panel review. No such regulation exists for technological pilots. While GDPR provides some protection in the European context, there is no equivalent framework in the majority world.

In Chapter 2 we discussed at length the lack of meaningful consent in refugee biometric registrations and digital identity systems where to refuse to submit one's biometric data amounts to a refusal to receive aid when there are no alternatives for survival. In the case of experimentation, the problem of consent takes the form of not making people aware of the nature of a programme and its potential limitations. As one interlocutor from the aid sector put it: 'the problem begins when not all people understand that they participate in an experiment'. If people are presented with a technological system that appears to be complete, and therefore safe to use, that has significant implications for their awareness of possible dangers. Even if people are presented with a consent form, if the nature of the pilot and its data management are not made abundantly clear then consent is not meaningful.

The catalogue of risks is in stark contrast to the humanitarian imperative 'do no harm' which stipulates that humanitarian organizations should not put affected people in harm's way as a result of their interventions.

Conclusion

The chapter began with a historical account which explains why people in the global South continue to be readily available as experiment subjects. In the chapter we traced the genealogy of experimentation from nineteenth-century experiments to contemporary pilots as part of the phenomenon of 'AI for good'. Most technological experiments involve partnerships between humanitarian organizations and private companies. An example is the *Building Blocks* partnership between the World Food Programme and IrisGuard, the biometric technology company that also specializes in border securitization. Technological experiments are occasionally solely commercial initiatives, usually framed as AI-for-good. Such practices turn private companies into de facto humanitarians, but without the imperative to adhere to the humanitarian principles.

The chapter observed that experimentation is a process of extraction. There are parallels with experimentation among marginalized groups who are often the first to experience technological innovations before these are rolled out more broadly. A case in point are prisons which, as Anne Kaun and Frank Stiernstedt have shown, have been used as a testing ground for technologies that are later adopted by the general public (2023). For example, the genealogy of popular technologies such as Fitbits can be traced to prison ankle monitors (Kaun and Stiernstedt, 2023). Eubanks (2017) has also observed how people living in poverty are the first to experience the digital welfare state before automation is rolled out to the general population. Parallel to extraction we observe an outsourcing of risk, especially in the case of experiments driven by technology companies. Should things go wrong, the risk is perceived to be lower if the experiment takes place in a distant, or othered land. The legal cost risks are certainly lower.

Experimentation in vulnerable settings can translate into publicity and, when private companies are involved, profit. The hype around chatbots, just like the hype around other innovations such as blockchain is one of the driving forces in the development and roll out of humanitarian innovations especially those involving the private sector. Humanitarian technologies also extract visibility and status which in turn can attract donor funding. Technological innovation is a way in which humanitarian organizations rebrand themselves as modern, agile and forward-looking.

By seeking problems for solutions, technological experiments are at odds with the humanitarian imperative 'do no harm'. Technological experiments reproduce power asymmetries by asserting Eurocentric values. Experimentation follows a North–South trajectory while projects often lack linguistic, or cultural sensitivity and impose instead imported categories on local contexts. The dominance of the English language in machine learning and artificial intelligence applications exemplifies the epistemic violence associated with chatbot and AI experiments. The absence of data management policies, clear accountability systems and informed consent further compounds the concerns. Because technologies are cast as enchanted objects, the power asymmetries of humanitarianism are heightened. The enchantment of technology also hides the work of mediation, and the power relations at stake.

Finally, the chapter charts a new type of experimentation, which I term surreptitious experimentation. Surreptitious experimentation becomes possible as digital infrastructures, such as biometric systems, become ubiquitous. This process of infrastructuring allows for a continuous flow of experimentation, which is not named as such, and which operates in the infrastructural background, and therefore remains hidden – yet in plain sight. Surreptitious experiments take place outside the laboratory, without clear boundaries, without meaningful consent, or processes of accountability. In so doing, surreptitious experiments compound the effects of extraction, coloniality and enchantment associated with other types of experiments. Surreptitious experimentation heightens the power asymmetries of humanitarianism and further marginalizes people who already face extreme precarity. In so doing, surreptitious experimentation encapsulates the most problematic form of technocolonialism.

5

The Humanitarian Machine

Automating Harm

When biometrics were first introduced by UNHCR in the repatriation programme of Afghan refugees in 2002 in order to address the problem of refugees making multiple claims, one of the officers stated: '*How can [refugees] argue now, the machine can't make a mistake.*'[1] I came across similar assertions during my fieldwork when automated decision making was deemed to be accurate, objective and impartial. Despite well-documented errors, the faith in processes of automation and computation remains undented. Recall that in Chapter 2 we estimated that up to 11,800 refugees out of a total of 396,000 in the 2002 response may have been erroneously excluded given the error rate in iris identification at the time.

This chapter aims to explicate the faith in the infallibility of computation and artificial intelligence. This is not a new development and is not exclusive to the humanitarian sector. It can be traced to the belief that numbers, and therefore processes of arithmetization, computation and automation carry authority. This is why statistics and automation are favoured by bureaucratic systems, which themselves are meant to inspire trust through processes of standardization. As we saw in Chapter 2, biometric technologies were introduced to eliminate doubts about refugees making multiple claims. Biometric registrations and authentification were instituted to address the suspicion of fraud. At the same time, biometrics and automation have also been introduced to address any doubts about humanitarian organizations. Digitization produces robust audit trails while artificial intelligence makes humanitarian agencies appear objective and neutral. By making decisions without appearing to make decisions, automation allows humanitarian organizations to claim to be apolitical. By helping them appear neutral and apolitical, digital innovation makes humanitarian organizations more attractive to donors. Through the adoption of digital practices, aid agencies appear more efficient and forward-looking. In turn, states also prefer the use of digital infrastructures which are interoperable with their own systems. Biometric technologies exemplify this trend. The private sector and technology companies in particular also favour the machine which offers profit opportunities, the testing of new technologies and the legitimation of operations through 'doing good'.

The humanitarian space is a fragmented space, governed by different logics. In this chapter I argue that the 'machine' is not just a metaphor to refer to AI, computation and digital innovation. The machine is also an actual material infrastructure that brings together the different actors and logics of the humanitarian field. The infrastructures of computation help achieve a semblance of coherence that brings together humanitarians, donors, states, entrepreneurs, and technology companies. Digital infrastructures and practices of computation build on the vast bureaucratic infrastructure of humanitarianism and take it to a new level. The notion of the machine does not imply that humanitarianism is a coherent field. Quite the contrary. The emphasis here is on the appearance of coherence, the importance of which should not be underestimated.

The chapter will start by outlining the bureaucratization of humanitarianism. Bureaucracies have been criticized for foregrounding their own internal priorities for survival, rather than the needs of the communities they serve (Ferguson, 1994; Gupta, 2012). Herzfeld (1992) has argued that bureaucracies produce indifference. The most searing critique of bureaucracy comes from Bauman, who has argued that bureaucracies dehumanize and obliterate accountability (Bauman, 1989). I will then turn my attention to the way in which digital infrastructures and computation shape humanitarian bureaucracy. Because digital technologies provide standardization, replicability, anonymity and portability they exemplify Weber's characteristics of bureaucracy. Digital technologies and artificial intelligence represent the logical next step in a long lineage of practices of standardization such as statistics. Digital infrastructures contain crises and turn them into manageable facts.

Automation and computation magnify the harms of bureaucracies identified by Bauman and others. Automated decision making can exclude and discriminate. The rigidity (binary thinking) of the machine produces errors with catastrophic consequences for minoritized people. Because digital systems are infrastructural, they rely on invisible and opaque processes which means that any errors are hard to redress.

Harms also occur at the everyday encounters with the machine: with the chatbot that doesn't include one's problem as part of the drop-down list of possible responses, or, simply, with the technology that doesn't work. Because it strips context out of its binary calculations, automation produces 'aphasia', the occlusion of knowledge and the difficulty of speaking about colonialism (Stoler, 2016). For these reasons, the digital infrastructure of digital bureaucracy produces structural violence.

While these critiques help explain much of what is wrong in humanitarian bureaucracies, they do not account for the care that also takes place in humanitarian settings. For all their faults, humanitarian organizations save lives and care for people in extreme situations of dispossession. How can we understand the paradoxical co-existence between care and indifference, and care and surveillance? Understanding this paradoxical relationship will also allow us to make sense of the very high degree of reflexivity and self-criticism that is present among humanitarians. Operating as a machine allows for the absorption of criticism without fundamental change.

Bureaucracy: A moral sleeping pill?

Humanitarianism in its contemporary iteration is a massive bureaucratic operation. Relief efforts depend on complex logistics about how to distribute aid to millions of people over vast territories. There are over 5,000 humanitarian organizations globally. This figure includes the eleven UN agencies specializing in relief assistance, 930 INGOs, approximately 3,900 national and local NGOs as well as the ICRC and IFRC and the national Red Cross and Red Crescent societies (ALNAP, 2022). Just to take one example, the World Food Programme, the UN's largest agency, had a budget of US $14.1 billion in 2022. That same year, the WFP distributed two million mt of food to more than 28 million people across the world. It also distributed over US $3 billion in cash assistance to people across 72 countries.[2] The bureaucratization and professionalization of humanitarianism are partly the result of this huge expansion involving complex logistics. Although theories of bureaucracy have been developed mainly in relation to states and governments, UN agencies and large NGOs share many characteristics with state bureaucracies.

There is a second compelling reason behind the bureaucratization of humanitarianism. In the aftermath of the Rwanda genocide in the 1990s, urgent questions were asked about the potential harms of aid. The sector's response was a further shift towards standardization and bureaucratization. Because bureaucracy standardizes and establishes objective and replicable processes in order to treat all citizens the same way, it is assumed to help humanitarian organizations achieve objectivity and impartiality. Ultimately, what is at stake is humanitarianism's apolitical status, which legitimates its capacity to intervene in conflicts and to operate within the sovereignty of nation-states.

Bureaucracy is characterized by hierarchical organization, division of labour and the primacy of efficiency and technical responsibility over moral responsibility. Humanitarian organizations are very hierarchical. Policy decisions are made at the headquarters, which are often based in cities in the global North. Like in all hierarchical organizations, humanitarian workers are accountable to their superiors. Like most bureaucracies, humanitarian organizations depend on the division of labour and the systematic separation of tasks. This is very evident in the UN 'cluster system', where each team is responsible for one area (be it shelter, or sanitation and hygiene) and nothing else. The criterion for evaluating technical responsibility is whether tasks are performed according to plan. The language of humanitarian organizations provides evidence of that. Terms like 'logframe' and 'targets' reveal that the priority is on ticking boxes on spreadsheets.

The most powerful critique of bureaucracy is put forward by Zygmunt Bauman (1989). According to Bauman, bureaucracy's hyper-rationality, impersonal nature, and obliteration of morality provided fertile ground for the Holocaust. According to Bauman, rational bureaucracy is 'an instrument to obliterate responsibility' (1989: 163). This is the result of the key elements of bureaucracy: the separation of tasks through the specialized division of labour and organizational hierarchy; the 'objective' conduct of business; the pursuit of rational calculation; and the application of rules to people without considering their humanity (Bauman, 1989: 106). As a result, Bauman argues, moral responsibility is substituted by technical responsibility, which is the evaluation of action against the targets set. This is a form of audit where, to paraphrase Bauman, 'humanitarian action is measured against itself' (1989: 98). Moral responsibility, on the other hand, involves action measured against its consequences. This is messier and more difficult to measure than technical responsibility. Accountability to affected people should be a typical example of moral responsibility. But, as we saw in Chapter 3, the bureaucratic and technologized performance of the feedback exercise turned what should have been moral into technical responsibility.

For Bauman, bureaucracy is a moral sleeping pill (1989: 26). Another equally important effect of bureaucratic activity is dehumanization: the possibility of referring to people in 'purely technical, ethically neutral' terms (1989: 102). We can see here that not only is the apolitical nature of humanitarianism problematic; it is, also, ultimately, deeply political. According to Bauman, dehumanization starts by distancing and by reducing people to quantitative categories; once dehumanized, people 'lose their distinctiveness' and are 'viewed with ethical indifference' (Bauman, 1989: 103).

Michael Herzfeld has also written about how bureaucracies, although they are meant to produce accountability, produce indifference (1992). Several studies have shown how bureaucratic systems tend to prioritize their own internal priorities for survival as opposed to those of their constituents (Ferguson, 1994; Gupta, 2012). The emphasis on rationality and indifference explains why bureaucracies have been likened to machines (Ferguson, 1994).

The counterargument in defence of bureaucracy is that without the impersonal application of standardized rules, the door is left open for corruption, nepotism and the furthering of inequalities. Paul du Gay has written in defence of bureaucracy, focusing on the importance of the 'ethos of the office' (2000). His argument offers a defence of the British civil service against managerialist critiques that push for more entrepreneurship in government (du Gay, 2000). While his argument certainly applies to current debates about the conduct of government in Western democracies and the undermining of the UK civil service by the populist right, his thesis is less applicable to the marketized humanitarian space, where the idea of the 'office' is not as well defined. The marketization of humanitarianism, with its constant pressures for funding renewal and the constant turnover of staff, undermines possibilities for staff accountability. With staff only staying in the field for a few weeks, before moving onto the next crisis, there is no 'office' so to speak. Ironically, in the absence of 'office' – where one officer oversees a policy and applies the rules systematically over a period of time – we encounter instead more audit, more paperwork, more data analytics and third-party monitoring. The marketized nature of the sector means that humanitarian officers are often stripped of their agency to actually perform their roles properly. The move to automated decision making in humanitarian operations is the logical next step for an industry that already undervalues its key workers.

Du Gay's intervention brings some nuance to Bauman and Herzfeld's blistering critiques of bureaucracy. Bauman's account does not help explain the reflexivity and thoughtfulness displayed by many humanitarian workers who are not devoid of the moral responsibility that he so laments (1989). The marketized bureaucracy of humanitarian organizations can help explicate why harms occur when the vast majority of humanitarians work hard, often under adverse conditions, to save lives. This chapter aims to account for the paradox of the coexistence of care and structural violence in the humanitarian sector. But before I do so, let's examine the way statistics, computation and AI complicate our understanding of bureaucratic operations.

The power of numbers and computation

Numbers and statistics carry authority which is why they have always been very appealing to bureaucracies (Hacking, 1990). Barnett observes that as early as 1931 'there was considerable emphasis and faith placed in the value of data and statistics' for relief organizations (2011: 93). As a technology of distancing, numbers derive authority from their capacity to create and overcome distance, both physical and social (Porter, 1995). Because the language of mathematics is highly structured and rule-bound, numbers can be easily transported into different cultural contexts. At the same time, numbers and statistics abstract from the human and the personal. As numbers are widely assumed to convey objectivity and validity, they are thought to confer legitimacy to the governments and bureaucracies which use them (Porter, 1995). Porter argues that the appeal of numbers is 'especially compelling to bureaucracies which lack the mandate of an election' (Porter, 1995: 8), something that is particularly relevant for aid organizations. Numbers and statistics are assumed to remove arbitrariness and bias from the system and give the impression of fairness and impartiality. The assumed objectivity of numbers 'lends authority to officials who have very little of their own' (Porter, 1995: 8).

If numbers and statistics are assumed to lend authority to humanitarian bureaucracies, then AI and algorithmic decision making are the logical next steps in bureaucratic governance. AI is routinely framed as scientific and, therefore, objective. Technologies are further celebrated for their efficiencies and the capacity to process large datasets and predict outcomes. Digital innovation and AI are not just championed as a tool that can correct the deficiencies of humanitarianism and restore objectivity, impartiality and participation; they are also imported to respond to increasing pressures for savings and efficiencies.

Even as early as 1989, Bauman predicted that advances in information technology would greatly increase the moral and psychological distance between bureaucrats and people and further diminish the relevance of moral responsibility. There are further concerns about statistics, computation and AI. Classifications – on which statistics and AI depend – are neither unambiguous, nor objective (Bowker and Star, 1999; Hacking, 2006). Datasets are inherently incomplete and reflect social biases about who or what is counted. Statistics make populations legible, which in turn makes populations manageable (Appadurai, 1993; Scott, 1998). Statistics are inherently political, in the sense that they are a tool for intervention (Asad, 1994: 76). Statistics have been used by colonial authorities, and later by postcolonial states, for the recognition, or, conversely, for the erasure of ethnic and linguistic majorities or minorities (Gupta, 2012: 158; Weitzberg, 2015).

Statistical methods do not just describe or count a population that already exists. As Ruppert and Scheel argue, statistics help to enact, to bring into being a population as a knowable object of government (2021: 4). Similar work is performed by AI applications, including data visualizations (Kennedy et al., 2016). Data practices do not just enact populations or other objects of bureaucracy, but also the bureaucrats themselves. Having made their populations – or other matters – legible, bureaucrats are then rendered as competent and knowable functionaries. People are also socialized to become 'that which can be measured', further entrenching classifications and making them even more invisible (Bowker and Star, 1999).

It becomes clear that the biases in data and AI systems are not the result of a glitch or inferior code that can be fixed. There will always be bias as long as classifications are the result of human decisions which are shaped by social, political and cultural contexts. Bias will persist given the inherently incomplete nature of datasets which reflect social and political orders and the inequalities they engender. Algorithms and AI are based on these classifications and the always-incomplete datasets. This is why the outcomes of automation and AI will always be shaped by the omissions, overrepresentations and miscategorizations of the system.

Humanitarian infrastructures

All this matters – not only in order to challenge the faith in objective AI and data in humanitarian bureaucracies; it also matters because digital technologies, data and AI are becoming increasingly infrastructural in the aid sector. In Chapter 2, we observed that biometric technologies and blockchain underpin a host of services and practices and bring together multiple stakeholders, from host and donor governments to global technology companies and local vendors. The interoperability of the systems also emphasizes the importance of making sense of these developments as an infrastructure from which data and data imaginaries flow.

In the book I have argued that the humanitarian space is a fragmented space, shaped by different logics according to the priorities of different actors. Humanitarian organizations are torn between their commitment to be accountable to crisis-affected people (logic of accountability), and the pressures of funding, which make them dependent on a small number of donors. Donors, who are largely national governments, are accountable to their citizens and their voters and demand evidence of where their funds have been spent (logic of audit). National governments are also driven by their own concerns regarding securitization and push humanitarian agencies to adopt technologies that allow for the reusability of humanitarian data for the purposes of state interests (logic of securitization). Private companies are driven by opportunities for profit, visibility and experimentation (logics of capitalism and techno-solutionism). It is impossible to summarize what people affected by crises want – this is hardly a homogenous group and so much depends on each empirical context. Judging from the Haiyan fieldwork, none of the issues mentioned in this paragraph would feature in my participants top one hundred priorities. In Chapter 6 we will observe how people in emergency settings appropriate digital technologies in ways that are meaningful to them.

In this fragmented field, digital technologies are one of the things in which all constituents are interested, albeit for different reasons. The concerted investment means that digital infrastructures are now in place and connect the different actors: humanitarians, donors, states, entrepreneurs, technology companies and people affected by emergencies. Biometric data, digital cash transfers, digitized feedback, health data, distribution data, satellite data, social media analytics, social media posts, are only some of the data that travel across the infrastructures that increasingly underpin the humanitarian space. I argue that the 'machine' is not just a metaphor to refer to the humanitarian bureaucracy. By 'humanitarian machine' I refer to an actual material infrastructure that connects all the 'stakeholders' in the humanitarian field. The notion of the machine does not imply that humanitarianism is a coherent field. The emphasis here is on the *semblance* of coherence.

The circulation of biometric data exemplifies the working of the 'humanitarian machine'. Examining the infrastructure allows us to see that the flow of data is either towards the state (securitization) or private companies (capitalism) – and not towards data subjects, who do not experience tangible benefits. In Chapter 4, we observed a similar direction of travel in the case of feedback data, which flow towards donors, thereby becoming audit data. Understanding the humanitarian space as an infrastructure on which data flow reveals the power geometries involved. It is not just data that circulate; the infrastructure also enables the flow of imaginaries about technology, ideas about participation and justice – to name a few.

We can discern that the boundaries between the humanitarian bureaucracy and state machinery are porous. The interoperability of systems – when they actually function – enables these synergies. This will become evident in the 'double biometric registration' example we'll explore later in the chapter. Yet it is important to recognize that I do not imply a convergence of all bureaucracies (state, commercial, humanitarian) into one. The infrastructures of computation enable flows and extraction between the constituent parts. Systems can be interoperable, but national governments and private companies can also retain autonomy.

In the rest of the chapter, I will draw the contours of the humanitarian machine by focusing on key examples.

Containing crises

A primary function of statistics is to make populations legible and, in turn, to enact bureaucrats as competent administrators. AI applications work in similar ways. To illustrate this, let's examine the World Food Programme (WFP) 'Hunger Map Live' platform, which is WFP's 'global hunger monitoring system' developed in partnership with the Alibaba Foundation.[3] As the name suggests, the platform is an interactive map of the world, which visualizes food insecurity in relation to climate or political events such as conflict. It combines 'AI, machine learning and data analytics from various sources, such as food security information, weather, population size, conflict, hazards, nutrition information and macro-economic data – to help assess, monitor and predict the magnitude and severity of hunger in real-time'.[4] The resulting analysis is displayed in the interactive map, which colour-codes areas with high food insecurity or other forms of precarity. According to the WFP website, the platform is aimed at WFP staff, key decision makers and the broader humanitarian community to make informed decisions relating to food security.[5]

'Hunger Map Live' was launched in 2019. During the first months of 2020 I interacted with the platform regularly, wanting to get first-hand experience of how it works and how it orders information. It became evident that the platform did not update as frequently as real-world events. Early 2020 of course, was marked by the emergence of the Covid-19 pandemic, when the world was gripped by the unknown nature of the new virus. As the death toll mounted and lockdowns were imposed, 'Hunger Map Live' was slow to update some of the dashboard data in March 2020.[6] Perhaps the events of early 2020 were too fast-changing for the AI platform, or perhaps states were slow to release relevant statistics. But, in any case, the promise of real-time data was not fully delivered during the first months of the platform's life. Despite the fact that the AI platform did not always provide 'live' data, it did perform important work.

World hunger is one of the most challenging problems the world is facing. The combination of statistics, AI and data visualization aim to make hunger a countable and therefore a manageable event. 'Hunger Map Live' uses the conventions of data visualization to acquire a sense of objectivity that enables it to do the persuasive work of visualizations (Kennedy et al., 2016: 723). For example, the two-dimensional map offers a bird's-eye view, while the inclusion of data sources and charts in the visualization makes a claim to transparency and authenticity (Kennedy et al., 2016: 731). As a data visualization, 'Hunger Map Live' produces knowledge and authority. At the same time, through the sleek visuals and interactive functionalities, 'Hunger Map Live' is a form of technological enchantment that hides the work of mediation. Critical cartography scholars have argued that 'maps do not just represent space and place, but help create them' (Cidell, 2008: 1208). Similarly, statistical methods do not just count, they also produce. Ruppert and Scheel (2021: 4) remind us that statistics are performative as 'they enact, rather than reflect populations'. 'Hunger Map Live' doesn't just represent or predict the reality – it renders the phenomenon of world hunger as a containable and measurable fact. Similarly, 'Hunger Map Live', produces the aid organization (WFP) as capable of managing the problem.

Statistics are so tied to bureaucracies that any policy goals are measured against the pre-defined statistical objectives. In Chapter 3 we saw that accountability and participation are measured against the criteria set out by relief agencies. Similarly, hunger will be reduced only when there are statistics to prove that it has been reduced. In other words, reducing hunger, entails reducing the statistics about hunger. AI platforms and data visualizations such as 'Hunger Map Live' contribute towards making the goal of 'zero hunger' – which is one of the UN's sustainable development goals (SDGs) for 2030 – appear more achievable. The application also makes the organization appear more capable of achieving the desired goal.

Apart from containing crises and making them look 'solvable', AI allows humanitarian organizations to make decisions – without appearing to make decisions.

The structural violence of automated decision making

One of the most pervasive encounters with the 'machine', involves the use of automated decision making in aid distributions. Because demand for aid outstrips supply, humanitarian organizations target households using different methods. A common approach is the use of the Proxy Means Testing (PMT), which categorizes households on the basis of characteristics which are then processed through machine learning algorithms to produce the distribution lists. The PMT method is controversial as it has led to problematic outcomes in a number of settings. A review of PMT distribution in the Kakuma refugee camp in northern Kenya found a 4.3 per cent exclusion error (Guyatt, Della Rosa and Spencer, 2016). Given that the camp hosts over 201,000 refugees, an exclusion error of 4.3 per cent equals 8,643 people.[7] A review of PMT targeting in Lebanon found that over 30,000 families excluded by the system appealed the decision.[8]

The lack of transparency about how scoring took place meant that automated decision making was perceived as arbitrary by those excluded. Frontline staff were equally unaware of the workings of the algorithm. A review of the system among Syrian refugees in Turkey, supported by the Danish Refugee Council (DRC), found that field staff were unaware of the workings of the system and could not explain targeting to those excluded. The opacity and unfairness of the system led to protests outside the DRC offices and on social media.[9] One of my interlocutors from the humanitarian sector described the use of targeting algorithms as 'random', making the targeting policy hard to explain to those who felt wronged by the system. According to my interlocutor, the secrecy around the PMT algorithm is intentional in order to prevent refugees from 'gaming the system'. For all these reasons, the Kakuma camp study recommended that 'blanket coverage of distributions is preferable as it complies with the "do no harm" principle' (Guyatt, Della Rosa and Spencer, 2016: 2).

Selective distributions are not new, of course, and they have always been controversial. In Chapter 3 we encountered the strong feelings of resentment harboured by those excluded from assistance during the Typhoon Haiyan response. In one of the examples discussed in Chapter 3, we saw that one of my interlocutors, Lia, had been excluded because of decisions made by the local community leader (*barangay* captain).

Recall that it is common practice for aid to be channelled to people via local government authorities, who may ultimately decide on eligibility criteria.[10] In Lia's case, the decision was unfair, but at least she could argue with the captain and the officials, who were her neighbours. When the decision is made by the machine without any clear explanation about the criteria used, then it is harder to challenge this outcome. That field staff are unable to cast light on the rationale for the decision only makes matters worse. At the same time the decision is framed as 'scientific' and therefore, 'objective' leaving little room for contestation. Even though computation and algorithms are essentially probabilities, the outcomes of automated decision making are treated as unambiguous.

Because AI is assumed to be scientific, objective, neutral and accurate, the machine is cast as superior. This accentuates the already unequal power geometries of aid. The more AI is anthropomorphized – the more it is given human characteristics, as evident in the phrase 'the machine thinks' – the more it dehumanizes actual people. While algorithms are known to include gender, race, ethnicity, class and disability biases (Benjamin, 2019; Noble, 2018), in this context no doubt is expressed about the reliability of the machine output. Machine learning biases are usually structurally determined. Because they depend on categorizations that exclude or misrepresent whole groups of people, they affect collectivities. Virginia Eubanks has shown how race and class intersect in the way automated welfare exclude the most marginalized citizens in the US (2017). In Chapter 2 we examined how biometric technologies produce race, gender and class because of the biases in the datasets that train artificial neural networks. It is striking that even though AI bias is structural, the burden of appealing a decision is individualized.

Automated decision making permeates all aspects of the migration and asylum management process. From the forecasting of migrant flows to resettlement decisions, machine learning algorithms inform a range of decisions (for a review see Ozkul, 2023). UNHCR's Project Jetson 'is a machine learning based experiment that provides predictions on the movements of displaced people'.[11] Accurate predictions based on publicly available data would allow the agency to be prepared for the arrival of refugees. One of my interlocutors expressed scepticism regarding the possibility and usefulness of such predictions:

> I was at this tech conference where they were discussing how we could have predicted the Syrian refugee crisis if we had better data analytics. Are they really serious? What could they have done? Build a wall to stop the refugees? Would that have prevented the crisis? [...] It quickly became clear that this would become a major crisis. Knowing a few days in advance would not have made a difference. Even one year after the first people arrived in Greece, there was no preparation. And this is because it's such a complex situation.

The use of artificial intelligence in the management of migration and asylum primarily involves nation-states and the technology companies that provide the border security infrastructures. It is, however, impossible to separate state practices out of our discussion as they shape refugee experiences. Crucially, digital infrastructures bring together humanitarian agencies, national governments and private companies. Infrastructures facilitate data sharing and blur the boundaries between the regimes of data governance. As we have already discussed, biometric data are a case in point – they travel within but also beyond the humanitarian sector, which creates vulnerabilities for data subjects.

AI is used widely in border control, exemplified by the use of biometric technologies. In line with the cultures of experimentation explored in Chapter 4, new AI infrastructures of border control are being tested. Of these, iBorderCtrl, has been the most controversial. iBorderCtrl, a pilot (2016–2019) based on emotion recognition technology, was essentially a 'lie-detecting machine', which aimed to identify whether travellers were lying. Other uses of AI include matching algorithms, where refugees are relocated on the basis of an algorithmic recommendation (Masso and Kasapoglu, 2020). Humanitarian organizations have also piloted relocation apps that aim 'to predict where resettled refugees are likely to thrive'.[12]

Many of these pilots have yielded poor results, leading to some being abandoned.[13] But, even as pilots or digital relics, the projects reveal a lot about the sociotechnical imaginaries behind AI, namely the fantasy that technology can control the future and make it manageable. Crucially, what is evident here is the replacing of moral with technical responsibility. The machine allows officers to make decisions without appearing to be making decisions. There are implications for accountability, too. If decisions regarding a refugee's resettlement are unsuccessful, who can be held accountable for this decision? The complex supply chains involved in the design, programming and running of AI programmes obliterate accountability.

Some of these questions are relevant in the following example, which focuses on the binary logic of the machine.

The binary logic of the machine

Structural violence is amplified by the rigidity of digital infrastructures and their binary logic. The case of the biometric registration in Kenya is a case in point. Between 1988 and 1991, thousands of Somali refugees arrived in Kenya fleeing civil war in their country. The large refugee camps in northern Kenya also attracted ethnic Somali Kenyan citizens, who had been struggling after a period of extended drought and who saw the camps as opportunities for accessing food and healthcare in the face of adversity. The problems emerged when the children of those Kenyans reached adulthood and sought to issue their first identity cards. Because their fingerprints were already registered in the refugee database, the Kenyan government refused to issue them with an identity card. This left an estimated 40,000 people under the age of forty in limbo. An identity card is vital for accessing all services in Kenya, from healthcare, opening a bank account or getting a job, to getting married and registering one's children.[14]

This example illustrates the potential harms of keeping biometric data indefinitely, which is the practice at UNHCR. If the biometric data had been deleted as soon as the refugees left the camp, then the risk of double registration might have been averted.[15] Of course, if data had already been shared with the Kenyan government, any later deletion by UNHCR would have been rendered meaningless. As Jacobsen points out 'when so much sharing takes place, it is hard to conceptualize how deletion can be effectively implemented' (2022: 642). When data are shared and permanently stored, biometric technologies turn the refugee camp into a diffused process, without clear boundaries. Encampment continues well after someone has left the camp. Biometric data challenge the idea of leaving the camp or exiting one's refugee status. If someone's biometric data will always remain in the vast refugee database of UNHCR, then at one level they will always remain refugees. The fact that these databases are interoperable and accessed by states (in this case the Kenyan government) further heightens the potential risks.

One could argue that in the double registration example, it was the Kenyan government that was responsible for the harm, as they refused to issue identity cards to those who had been registered as refugees by UN agencies. Even if humanitarian organizations are not directly responsible for the exclusion of Kenyan nationals, their predicament cannot be understood outside the digital infrastructure of humanitarian bureaucracy. The biometric registrations originated in the encounter between refugees and UNHCR. If data collected by UNHCR later become the cause of harm, UNHCR should bear responsibility. If the imperative is to 'do no harm', then the humanitarian organization bears responsibility, whether the re-use of data is intentional (sharing agreements with governments), or unintentional (data breach). The point is that state and humanitarian bureaucracies intersect, and the digital, interoperable systems facilitate that. The immutability of biometric data and the 'for ever' data retention practice of UN agencies further compound the risks to affected people.

This example also brings to light the binary logic of biometric registrations. The machine can only offer two options: either one's biometric data classify them as a refugee or as a Kenyan national. The binary logic does not reflect the complexity of human life, which is always a lot messier than the machine allows. This is more than just a matter of inaccuracy or lack of nuance; it is about the violence of being defined as someone who is not you, as Édouard Glissant has put it (1997). This violence is ultimately dehumanizing, as it strips away people's agency to define themselves and claim their rights. The testimonies of those affected by this machine error are poignantly captured in an investigation and video documentaries by Privacy International.[16] Just like the earlier examples discussed in this chapter, the onus was on the young Kenyans to individually prove their citizenship status. The machine did not have a clear pathway of redress. Perhaps it was considered unthinkable that the machine would ever make a mistake. What comes across clearly from the Privacy International investigation is that the process of undoing a 'machine error' is long and arduous. The machine takes away people's agency to define themselves and then places all the burden on them to correct the error and prove the legitimacy of their existence.

The double registration example also brings to the surface the way the data are implicated in the colonial legacies of contemporary bureaucracies. As is often the case in colonial regimes, in Kenya there were poor records about minoritized people such as Somali Kenyans. Keren Weitzberg observes that Kenyan leaders 'inherited long-standing practices laid down by the colonial state, which was unable to transform its subjects into a countable, traceable population' (2015: 411). Given the incomplete records, it was particularly difficult for the Somali Kenyans affected by the double registration to prove their citizenship. What also becomes apparent here is how the postcolonial state internalizes the practices of the colonizers – which has been a theme throughout this book. In response to their erasure, some of those affected by the double registration developed tactics to bypass the rigid system by forging identity documents. As Weitzberg observes, forgery in this case was more truthful to people's identities than the bureaucratic or algorithmic errors (2020).

Affective encounters with the machine

The rigidity of the machine is felt not only in the violence of misrecognition, but also in the everyday encounters with the digitized bureaucracy. The most common harms are experienced in an ordinary interaction with a chatbot that doesn't work, or in not receiving a response to a complaint submitted via a messaging app. In Chapter 3 we observed how humanitarian 'accountability to affected people' programmes prioritize certain forms of written feedback through digital channels. For those who overcame cultural barriers to submit their concerns, to not receive a response was demoralizing and disheartening. A common grievance among those who submitted feedback was that they were excluded from the aid distributions, which was hard enough. Not being listened to when feeling wronged adds insult to injury. Dolores, whom we met in Chapter 3, was clear that she would never use a feedback system again.

The text-based character of feedback channels is another way in which the machine excludes. Most bureaucracies prioritize written complaints and writing has been analysed as a modality that further marginalizes people living in poverty (Gupta, 2012). One of my interlocutors from the 'digital humanitarianism project' was very frank about the inequities of text-based participation. Anna worked in an NGO in Greece:

> We hardly hear from women online. [...] So that's definitely a limitation of communicating with people online that we've noticed, we don't hear women's voices, and of course we don't hear illiterate people's voices. And I would imagine it's often the most vulnerable people who don't, who aren't able to communicate with us online. So again it just reinforces the fact that face-to-face is so important.

The machine presumes a specific kind of 'beneficiary', who can write, who can summarize their problems succinctly and who is adept at using technologies, as we'll see below. In other words, the machine expects people affected by crisis to be like 'Western' people. The automation of bureaucracy becomes a vector of the coloniality of power (Quijano, 2000), an 'epistemic machinery' through which Eurocentric knowledge is reproduced (Knorr-Cetina, 1999). The expectation that the subaltern can only gain recognition through the categories of the West exemplifies Spivak's notion of 'epistemic violence' (2010). This is evident in the following example of an intervention aimed at the 'unbanked' following Typhoon Haiyan. Unbanked refers to people without bank accounts, which is considered as an indicator of 'economic development'.

During the Haiyan fieldwork one NGO together with a microsavings bank developed a cash assistance programme aiming to help local people open their first bank account. Those selected based on government criteria, were asked to open a mobile savings account and were then given SIM and ATM cards linked to the account. Some of our participants complained that even though they received messages that their funds had arrived, when they went to withdraw their funds, there was nothing there. One participant, Bato, described how he travelled a long distance to get to the closest shop with an ATM. Bato initially thought he did something wrong. 'Perhaps I didn't understand the message', he told me. It was only after four half-hour journeys that the money was actually transferred.

This example is indicative of how the system aspires to produce rational subjectivities with bank accounts and savings, but also ends up producing feelings of inadequacy in front of the machine. That the system is designed in the North and implemented in the South, in an area with no relevant infrastructure even before the Typhoon – let alone after one of the strongest storms ever to make landfall – reveals the coloniality of the approach. Introducing banking top-down is an example of epistemic violence. Introducing banking and the value of savings in aftermath of an emergency, when people had urgent needs for cash to rebuild their homes and their livelihoods, displays not only paternalism, but also a lack of understanding and sensitivity. The fact that the system was rolled out in a rural area without any infrastructure in place – few shops connected to the ATM system while there were no banks – could even be interpreted as cruel. It is a plausible assumption that what the microsavings bank aimed at was to register new customers, who would then become regular customers following the pilot. The instrumentality and potential profitability of the experiment make this a particularly disturbing encounter with the machine.

In this case the infrastructure behind the machine did not work initially. Even as the infrastructure failed it produced affect and subjectivity. Bato expressed feelings of inferiority and self-doubt. These were the result of the 'epistemic gaze' of the machine that othered Bato and positioned him in a position of inequity. The notion of the 'epistemic gaze' combines Fanon's notion of epidermalization, which refers to the internalization of inferiority, following the gaze of the other and Spivak's notion of epistemic violence. Just as the encounter with the biometric scanner is a form of 'digital epidermalization' (Browne, 2015), the encounter with the digital infrastructures can produce subjectivity. Infrastructures form subjects not just at a technopolitical level, but also through 'the mobilization of affect' (Larkin, 2013: 333). The 'epistemic machinery' of innovation compounds the existing inequities between affected people and those who provide aid.

Having outlined some of the harms of the machine, we should not lose sight of the fact that the machine – with all its significant faults – does save lives. There is a tension and contradiction here between the intentions of many individual humanitarian workers and the actual consequences of the machine for people. How can a sector that aims to do good, produce structural violence? How can extremely hard-working and well-intentioned individuals, end up contributing to a machine that reproduces coloniality?

'The change that doesn't change anything'

During my ten years of research in the sector, I came across some incredibly thoughtful and dedicated humanitarians, who made astute observations and critiques about the shortcomings of humanitarianism and the harms of technological experimentation. My own thinking has been shaped by these conversations. Humanitarians on the whole are very self-reflexive and self-critical. I have often been surprised by the openness displayed by practitioners in response to some of the arguments developed in this book. Rather than being shut off, I have been invited to several meetings, workshops and conferences to share my views and discuss how 'to do better next time'.

Given that there are strong critical voices within the sector, the question is why is reform so difficult to achieve in the humanitarian sector? We have already explored in Chapter 1 the various attempts to address long-standing problems in the humanitarian field. The response to each crisis in the field has been to introduce more bureaucracy. The aftermath of the Rwanda genocide caused a lot of soul-searching and recognition of where things had gone wrong. But the response was to standardize, professionalize and bureaucratize – a trend that has continued since and which seems to be accelerating with digitization and automation. Humanitarianism responded to critiques by asserting its apolitical, neutral, impartial, independent and objective character. Bureaucracy was deemed the method of achieving reform. Not only is bureaucracy not radical enough, it accentuates existing issues and even engenders new ones. Because the structural asymmetry between 'aid givers' and 'aid receivers' is not challenged, timid policy shifts have not addressed the fundamental critiques of power and paternalism. Ironically, some of the key reform policies of the last thirty years may have created new problems.

One policy merits some discussion. Localization entails increasing local actors' involvement in humanitarian operations in order to address the structural inequities within the aid sector. Localization was enshrined in the 'Grand Bargain' agreement that emerged at the World Humanitarian Summit of 2016. Localization policies involve (1) increasing the international aid funding channelled to 'local' NGOs to 25 per cent by 2020 (a goal that was not met); (2) the development of partnerships, and

(3) 'capacity building'.[17] The problem with localization is that it usually translates into hiring local staff during an emergency response. Because the policies are still decided in the European headquarters of organizations with 'local staff' expected to implement them, then localization does not challenge the power dynamics of humanitarianism. Similarly, if 'capacity building' means that local NGOs should emulate international ones, then again, the whole purpose of localization is rendered meaningless. As a top-down notion, 'capacity building' assumes local actors' 'lack of capacity' instead of recognizing the value of local approaches and knowledge. In addition, funding barriers further complicate the situation. Donors (national governments) are nervous about whether local NGOs can deliver clear audit trails which are deemed necessary to justify the overseas development assistance budget to their taxpayers.

Humanitarianism is not a homogenous field. It is also very hierarchical. Having spoken with frontline staff, many of whom come from the affected regions, and more senior officers between 2014 and 2021, the divergence of views is often striking. Frontline staff, tasked with saving lives, aid distributions, livelihood programmes and so much more were incredibly reflexive and thoughtful interlocutors. At the same time, they were also acutely aware of their limited power. Their room to manoeuvre was very limited and the bureaucracy ensured that they did not deviate from the task – even when they may have had misgivings about a particular policy. Although there were some notable exceptions, those higher in the organizational hierarchy often had an abstract understanding of the circumstances on the ground. Yet it was the more senior managers who drove policy (for similar observations see Slim, 2015).

All this is further complicated by the arrival of new actors in the humanitarian space. As we have seen throughout the book, the humanitarian space is a very fragmented one. Among the different groups, the latest arrival consists of the technology companies and entrepreneurs, who are driven by the logics of capitalism and solutionism. This group is often reluctant to adopt a reflexive approach that questions the workings of the machine and its potential harms. As the participant from the 'digital humanitarianism project' put it:

> I was at a conference recently and there were two main groups. The tech people, represented by big players, and the humanitarian people. Ten years

> ago, there was a group of us who were in-between, at the intersection between tech and humanitarianism. This group is not as strong [any more]. The tech people don't understand humanitarian issues. Tech companies are there because they see market opportunities. There is also this increasing awareness that tech companies need to show they are doing good in the world. They need to be ethical. Being involved in humanitarian activities is a way of showing that. It is a good branding strategy.

One of the greatest barriers to change is the presence of a bureaucratic machinery, which, like all bureaucracies, prioritizes its own goals. Bureaucracy – with its goals, targets and standards – depoliticizes reforms. Localization policies provide a good example. Rather than starting conversations with local agencies about building equitable partnerships, policies focus on metrics, and therefore, what can be measured. In so doing, humanitarian organizations are not measuring localization, but the targets they have set for themselves. In other words, 'localization targets' measure international organizations – not local agencies – and ultimately serve the donors' agendas. By expecting local NGOs to implement top-down policies and to then report following cumbersome bureaucratic protocols, international aid agencies devalue local knowledge and reproduce the asymmetries of humanitarianism. As one of my interviewees from the 'digital humanitarianism project' put it when describing the data demands of international partners: 'We are constantly feeding the beast.' Rather than treating local NGOs as equal partners, localization policies impose a Eurocentric model of aid and a stifling bureaucracy, which entrenches the disparities within the aid sector.

Contemporary humanitarianism involves complex supply chains involving donors, partners, vendors, local staff, technology companies and platform developers. In the case of UN agencies or large INGOs these networks can be labyrinthine. The division of labour and the separation of tasks, compounded by organizational hierarchies and increasing professionalization, diffuse and obliterate moral responsibility. The turning of complex issues into matters that are addressed with technical solutions further removes workers' agency to effect change. The presence of newcomers, such as private-sector companies and technology entrepreneurs, who have their own agendas, further complicates efforts to reform.

The humanitarian field has a great capacity to be open to criticism, absorb it, make some tweaks (usually involving more bureaucracy and standardization) and then emerge strengthened (de Waal, 1997: xvi). Individual critiques or acts of courage cannot change the system. Humanitarian reform often resembles 'the change that brings about no change' to paraphrase Angela Davis (in her interview to Gary Younge, 2007). The notion of the humanitarian machine explains why despite the efforts of care, the machine ends up producing structural violence.

Conclusion

The chapter has argued that digitization and automation magnify the limitations of humanitarian bureaucracy and end up producing harm. Yet the faith in technology and computation remains strong. The chapter has traced this appeal to the imaginaries of numbers and statistics as objective. The appeal of numbers, and by extension, of AI and automation, is that it confers legitimacy to bureaucracies. In their quest to appear neutral, impartial and apolitical, humanitarian organizations have turned to bureaucracy with its emphasis on standardization and replication. AI and automation are assumed to further remove bias from the system and give the impression of fairness and impartiality. AI predictive models and visualizations contain crises and make them appear manageable. Automated decision making, which is being introduced in humanitarian operations, allows humanitarian organizations to make decisions without appearing to make decisions.

Yet the reality could not be more different. Because technologies and infrastructures are political, they do not achieve the desired goals of neutrality and impartiality. Because AI depends on subjective classifications, its decisions reproduce bias and exclusion. As decisions regarding aid eligibility are outsourced to algorithmic systems and automated decision making, errors are likely to become systemic. Even though classification errors are likely to affect groups of people, exclusions are often experienced in an individualized way. The onus is always on the individual to fight their case and seek redress.

The binary logic of the machine does not capture the messiness of everyday life. It imposes categories and classifications even when these do not reflect people's lived experiences. Processes of automation assume a certain subject who can read and write and express their complaints in a concise way. Automation imposes a prototypical 'Western' subject on refugees or on disaster affected people which is a form of epistemic violence. The 'sanctioned ignorance' (Spivak, 1999) of the machine is evident when projects that assume the presence of technological infrastructure and the needs of the local population are rolled out without first bothering to find out whether an area has ATMs or whether people need savings accounts when they have far more urgent needs like shelter.

The infrastructural character of digital systems means that they underpin basic aid operations. Infrastructuring also implies that humanitarian systems are increasingly interoperable with those of nation-states and, often, those of private companies. The data sharing agreements between relief organizations and governments coupled with the permanence of records amplify risks to individuals as we saw in the case of the double registration in Kenya. Encampment turns into a permanent state: even when someone leaves the camp, they will always be traced back through their permanent records. All these observations exemplify the ways in which technocolonialism reworks and revitalizes the colonial genealogies of humanitarianism and of technology.

The chapter has argued that the infrastructuring of humanitarianism offers a semblance of coherence in what is a fragmented field. As digital systems and networks begin to underpin essential operations, the infrastructuring of humanitarianism resembles a machine. The metaphor of the machine is often associated with bureaucracies. The difference here is that this machine is not just a metaphor. It has a material dimension as the socio-technical assemblage of technologies, platforms, software, AI, blockchain, cloud computing, devices and cables as well as people in need, agencies, donors, private companies and their associated practices.

The notion of the machine helps explain why the labour of so many humanitarians, who dedicate themselves to saving lives, often ends up producing structural violence. The machine has its own logic dictated by bureaucratic hierarchies and the pressures of marketization. Despite the reflexivity and critique from within the sector, the machine is able to absorb the criticism and continue to operate as normal.

We will return to the notion of infrastructuring and its consequences in the Conclusion. But first, we'll delve into the micro level of everyday life which reveals that infrastructures and technologies are inherently contested.

6

Mundane Resistance

Contesting Technocolonialism in Everyday Life

In the days after Super Typhoon Haiyan made landfall, humanitarian radio stations were set up in the islands of Leyte and Samar as part of the response efforts. All local media, from newspapers to television and local radio stations, were out of service as buildings and infrastructure were destroyed. The city of Tacloban and the surrounding areas on the island of Leyte were without landline, mobile phone or internet connectivity for several weeks.

Some of these emergency radio stations only lasted for a few months; others operated for over a year, even after the commercial radio stations began broadcasting again. Humanitarian radio stations were set up with the primary aim of providing information for affected communities. In the first few weeks, programming focused on how to register missing persons or how to access clean water, medications, and petrol. Gradually, announcements included advice regarding sanitation and hygiene, information about the distribution of aid or the availability of services. Another important function of humanitarian radio stations was to encourage listeners to submit their feedback, usually via SMS through the station's dedicated hotline. At least one station used the open-source platform FrontlineSMS, which enabled listeners to send SMS texts to a designated number at a very low cost. FrontlineSMS then aggregated these short messages into databases, enabling the processing of feedback. These feedback hotlines were part of the accountability to affected people initiatives that we have explored in Chapter 3.

Humanitarian radio is very different from commercial radio. There are no intervals for advertisement and no background music. The tone is set by the circumstances, which in Tacloban were exceptionally tough by any account. Yet several of the humanitarian stations included slots with music programmes. One radio station included a music show every Sunday, where the local presenter would bring his guitar and play songs requested by listeners via the feedback hotline. Friends and colleagues of the presenter often joined in the studio and sang along to the music on air.

When perusing the station's feedback database, it was clear that the majority of the messages were requests for songs in anticipation of the music slot. In fact, most messages were song dedications to the listeners' relatives and friends. 'I am dedicating the song to my eldest sister Jessa who is celebrating her birthday today, stay strong', was a typical message. Comparing the radio feedback database to others like the ones described in Chapter 3, I was touched by the humanity of the messages and the stories hidden behind the dedications. Listeners were trying to rebuild the bonds that had been so brutally disrupted by the destruction caused by the storm surge. With over 6,300 dead and thousands more missing – and millions displaced – people were trying to rebuild not only their homes, but also their relationships. At the same time, music in and by itself was a way of transporting people back to a situation of ordinariness, to the rhythms of everyday life before the storm.

Listeners appropriated humanitarian radio and 'feedback' in a way that was meaningful to them. As is often the case, technologies are invested with meanings that may diverge from the designers' intended uses. In Chapter 3 we observed the problematic practices associated with digital feedback. Here we see that local communities appropriated a digital feedback platform in unanticipated ways. By turning a platform intended for information dissemination into a song and dedication playlist, listeners challenged the epistemic frameworks of the feedback mechanisms. Such practices can be understood as small acts of resistance by local communities. This chapter theorizes these practices as a form of 'mundane resistance'.

Technocolonialism is not a monolithic force imposed on people. It is actively contested and even resisted. We have already encountered several examples of resistance in previous chapters. The book began with the Rohingya protests against their biometric registration and digital identity cards. Although those protests ended with violent clashes and the continuation of the biometric registrations, they were an example of overt resistance. Not all forms of protest are as explicit and forthright. This chapter will primarily trace the resistance that takes place below the radar, and which takes different forms: from refusing to engage with digital practices such as digital feedback platforms, to appropriating technologies in unexpected ways as in the case of humanitarian radio. I term these practices 'mundane resistance' to emphasize its expression through everyday, ordinary acts. In so doing I will argue that techno-colonialism is inherently contested.

The literature on social movements and resistance has largely favoured the imaginary of revolution and has prioritized the study of direct activism. Yet outright protest is not an option in many parts of the world, where the consequences of speaking up can be prohibitive. There is a Western bias in the literature of social movements and protest. While dissent is largely tolerated in liberal societies, for people living under oppressive regimes, or in very asymmetrical situations, open protest can put their lives or their welfare at risk. Of course, protests do take place in authoritarian contexts (Sreberny and Khiabany, 2010) and often culminate in uprisings, as happened with the so-called Arab Spring (Kraidy, 2016; Tufekci, 2017).[1] But, if we look beyond the relatively few spectacular eruptions of protest, if we focus below the surface, we can discern a myriad of practices that also constitute small acts of resistance. These mundane and latent acts of resistance are often a necessary preparation for the expression of more direct forms of activism.

To make sense of the forms of mundane resistance, I draw on colonial and decolonial theory and, in particular, the Black radical tradition. I will also examine parallel debates in the fields of migration research as well as media and internet studies regarding the role of technology and activism and power and resistance. The chapter will focus on three cases of activism: direct activism, digital witnessing and mundane resistance. The latter is illustrated through the refusal to use digital technologies, the appropriation of humanitarian technologies in unexpected ways, and the uses of social media infrastructures for mourning. The chapter will end with a reflection about whether mundane resistance constitutes social change, or whether it reproduces power relations.

Resistance in asymmetrical settings

Resistance in conditions of extreme asymmetrical power relations may be limited by structural factors, but it does take place. Colonialism has always been a struggle of resistance (Fanon, 1961; James, 1938; Said, 1994). The Haitian revolution, which started in 1791 and culminated with independence in 1804, is an exemplar of resistance to colonial rule, as well as, according to C. L. R. James, a new paradigm of revolution (James, 1938). Priyamvada Gopal has examined how anticolonial struggle in the periphery of the British Empire was key for the emergence of dissent in the metropole (Gopal, 2019). For Frantz Fanon, decolonization is a process of revolution to change the order of the world (1961). Resistance need not only be armed or political; it is also symbolic. According to Edward Said, novels became 'the method colonized people use to assert their identity and the existence of their own history' (1994: xv).

In order to make sense of resistance in asymmetrical settings I draw on writers from the Black radical tradition, authors like C. L. R. James, Orlando Patterson, Cedric Robinson and Anton de Kom. Small acts often pave the way to the spectacular eruption of revolutionary fervour. When writing about the slave uprisings that led to the Haitian revolution, James recognized the role of folk culture and, in particular, of music and religious practices in the resistance process (1938). De Kom's account of resistance against slavery in Suriname focuses on the maroons, the escapees from the plantations who fought the colonizers over several years (2022). Patterson and Robinson observed that in very asymmetrical settings, such as slavery, where acts of outright defiance are not possible, resistance often took passive forms, including deliberate evasion, refusal to go to work, or satire (Patterson, 2022; Robinson, 2021). Patterson distinguishes between two types of resistance to slavery: violent and passive. 'Passive resistance' includes the refusal to work, general inefficiency, deliberate laziness or evasion, satire and singing, among others (Patterson, 2022: 261).

James Scott (1985) drew from Patterson's notion of passive resistance (Patterson, 2022: 6) when developing the notion of the 'weapons of the weak' to describe how everyday forms of resistance among marginalized people operate 'below the radar', including practices such as foot dragging, feigned ignorance, boycotts, as well as sabotage. In order to unearth these 'below the radar' practices, Scott develops the notion of 'infrapolitics' or the 'hidden transcripts': the opportunity to hear the accounts of marginalized people (Scott, 1990).[2] The notion of the 'hidden transcripts' contains a critique of power and is particularly apposite for the asymmetrical relations such as those we encounter in humanitarian settings. Scott (1985) drew on his own ethnography with peasant groups in Indonesia, who engaged in class struggle with those who were extracting labour, taxes and interest from them. While Scott's interlocutors avoided any direct confrontation with authority, they engaged in acts of pilfering, sabotage, arson, feigned ignorance, slander and subterfuge. Scott argues that these acts need to be understood as forms of resistance which can have cumulative effects in the long term.

Making sense of the 'weapons of the weak' presents us with a methodological challenge, given that the questioning of authority operates 'below the radar' (Scott, 1985). Scott, influenced by Goffman's dramaturgical model (1990), develops the notion of the 'hidden transcripts' to refer to the practices of resistance that take place in the 'back region'. Hidden transcripts need to be understood in relation to open transcripts, a term that refers to what is said by subaltern groups publicly, or, to use Goffman's terminology, in the 'front region', when oppressors are present. Hidden transcripts refer to what is said behind the scenes, when the ruling classes or oppressors aren't present, and subaltern groups are free to express their critical views. To listen to the hidden transcripts requires the researcher to 'descend into the ordinary' (Das, 2007). It is only through an ethnographic encounter that researchers can gain the trust of their participants and notice these small acts of contestation.

Within migration research, the 'autonomy of migration' approach offers important insights into human agency and resistance. The autonomy of migration approach foregrounds human mobility over state securitization. Border controls and securitization are a *response* to migrants' agency and not the other way around (Papadopoulos and Tsianos, 2013; Scheel, 2019). Migrants circumvent control and create social change by becoming imperceptible. According to Papadopoulos, Stephenson and Tsianos becoming imperceptible is an act of resistance: 'the most effective tool migrants employ to oppose the individualizing, quantifying, policing and representational' powers of north Atlantic states (2008: 217). Papadopoulos and Tsianos (2013) analyse migrant agency as a form of 'social non-movement', a term that Bayat originally proposed to capture the discreet and sustained ways in which subaltern groups in Middle Eastern societies struggle to survive by asserting their 'right to city' (Bayat, 2013: 15–16). Such practices can include squatting as well as a range of invisible everyday practices, which often pave the way for more radical mobilizations. This is a different form of politics, which operates below the radar of recognized political structures. To make sense of this agency, we must again delve into the everyday lives of subaltern groups, migrants or those affected by disaster.

In developing the notion of 'mundane resistance', I also draw on debates in media and internet studies. The study of social movements illustrates the parallel and contradictory potentialities of networked technologies (Treré and Bonini, 2022). Social media and communications infrastructures have afforded visibility to social movements and facilitated the micro coordination of protestors. Yet, at the same time, networked technologies have been used to monitor and control protest movements (Fenton, 2016). Processes of datafication and automation have had the most adverse impact on marginalized and racialized communities (Benjamin, 2019; Eubanks, 2017). At the same time, indigenous data sovereignty movements offer opportunities for visibility and social justice (Kukutai and Taylor, 2016). Technologies and infrastructures are tools of control, but also tools for resistance (Cowen, 2020). The contradictory consequences of media technologies and infrastructures were crystallized during the Covid-19 pandemic. In the lockdowns that much of the planet experienced following the public-health response to the pandemic, digital technologies became at once the means through which social life unfolded, and the means for surveillance and control (Madianou, 2022).

It is clear that media technologies and infrastructures do not determine outcomes; technologies shape society and are equally shaped by society and politics. Further, acknowledging the power of media corporations and media infrastructures is not incompatible with recognizing the agency of people to appropriate technologies in meaningful ways. There is a tendency in media and internet studies to favour a binary approach: either media technologies are powerful, or people have agency. I here argue that we should not make sense of the media power and human agency relationship as 'either/or', but as 'and'.

I develop the notion of 'mundane resistance' in order to draw emphasis on the everyday nature of resistance. The term mundane captures the everyday character of resistance, while also indicating that this is an active process. I prefer the term 'mundane' to Patterson's 'passive resistance' (2022), as there is nothing passive about refusing to work or going slow. Just like 'infrapolitics' (Scott, 1990), mundane resistance operates below the radar and is often latent. Mundane resistance operates on a spectrum with open confrontation and protest at the other end. To capture the movement along the spectrum the chapter will explore three moments of resistance: protest and activism; digital witnessing and storytelling; and mundane resistance.

Protest and activism

Resistance takes place on a spectrum between mundane acts on the one end, and direct activism on the other. While most of the chapter is dedicated to outlining the contours of mundane resistance, I acknowledge that open protest and confrontation do take place in situations of power asymmetries. We have already encountered the Rohingya protests earlier in the book. Despite the structural factors that suppress mobilization, protest in refugee camps is not uncommon. The Moria refugee camp on the Greek island of Lesvos was notorious for its overcrowding and abject conditions before it completely burnt down in September 2020. At that time the camp hosted over 13,000 people, which was more than four times over its original capacity of 3,000.[3] There had been several protests in Moria between 2016 and 2020, with refugees demanding from the Greek government better conditions and their transfer to mainland Greece. Similarly, refugees organized dozens of demonstrations, sit-ins, and roadblocks in the Za'atari camp in Jordan in 2013, culminating in violent protests in April 2014 (Clarke, 2018). Ethnographies of encamped life show that refugees engage in protest and activism to achieve visibility and an improvement to their conditions (Agier, 2011; Feldman, 2008; Malkki, 1995).

Although both the Moria and Za'atari examples involve protests, they are very different because the sociopolitical contexts behind each camp vary widely. In the case of Moria, the host country (Greece) was responsible for managing the camp. In most other cases (such as Za'atari) camps are co-managed by UNHCR and host governments. For example, Za'atari is jointly managed by UNHRC and the Syrian Refugee Affairs Directorate (SRAD), a Jordanian government agency.[4] We observe again the close relationship between humanitarian (especially UN) agencies and national governments, which we have already encountered several times in the book. The political context matters, as camp governance shapes the conditions refugees experience as well as the possibilities for collective action and the response to that action (see also Clarke, 2018). A further observation is that the high concentration of people in the confined space of a camp, is a contributing factor for the eruption of protest. This becomes evident when comparing the occurrence of protest among refugees in camps and among those living outside camps, where protest is less frequent (see Clarke, 2018 for a comparison among Syrian refugees in Za'atari camp and Syrians in Lebanon and Turkey; Malkki, 1995).

The conditions are altogether different in the aftermath of disasters, such as Typhoon Haiyan. Unlike the situation in refugee camps, the response to disasters does not involve high levels of securitization. As is typical in disaster emergencies, in the aftermath of Haiyan the national government was primarily responsible for the recovery. The Philippine government had set up a new government agency, the Office of the Presidential Assistant for Rehabilitation and Recovery (OPARR), with the explicit purpose of coordinating the relief efforts in the affected areas. Also involved in the recovery were the local government departments, which often had different priorities and different understandings of the local issues from those of the national government. Humanitarian agencies operated at the invitation of the Philippine government; something they were keen to remind me of on numerous occasions. These levels of governance matter if we are to understand people's acts of resistance. Resistance and activism are relational – so we must understand what is being resisted, and in relation to whom.

In the aftermath of the Typhoon Haiyan fieldwork, we encountered one example of grassroots activism. This was a rare example, but confirmation that vocal resistance can take place under certain conditions. In the first *Harampang* (community consultation), which I attended in Tacloban in June 2014, I was struck by the fact that it was mostly women who stood up to ask questions and speak about their unlivable conditions. One of those women was Dina, a 26-year-old from one of the worst affected *barangays* in the city of Tacloban. Dina had lost several of her relatives and friends, her home and her job. Dina stood up during the consultation and spoke eloquently in front of more than 200 people. She explained that the conditions in her *barangay* were unbearable. She complained about why aid took so long to arrive in her neighbourhood.

Dina lived in an old tent with her young daughter and other relatives. The tent was given to her days after the Typhoon made landfall. It was already worn out, but it became completely unusable by the time of the community consultation seven months later. Water came through the roof every time it rained. The government had promised new tarpaulin to be distributed in March – but, by early June, Dina and her neighbour Carol were still waiting. Together with other women, Dina and Carol started an SMS campaign, essentially sending a barrage of text messages to the local government representative. Carol recalls her SMS as follows:

> Good morning, Sir. We are from *Barangay* Lido, and we are in need of tarpaulins, so we would like to request from you. We just want to ask for updates on when we will get them, because our roofs are leaking, our situation is hard, and it is hot.[5]

Carol and Dina called this barrage of SMS 'text brigade'. Their efforts succeeded and tarpaulins were delivered within days. Carol said that the text brigade was a way for the officials to understand 'the feelings of those who were affected by the typhoon. We really need to speak out so they will know our concerns and the kind of hardship that we are experiencing so they can help us.'

Carol and Dina were supported by a Manila-based advocacy group, who are dedicated to empowering people living in extreme poverty in informal settlements. Community workers associated with the group arrived shortly after the storm and lived in Carol and Dina's neighbourhood for months. They provided mentoring and help with practical matters, including house building skills. One community worker in particular was pivotal for mentoring the women. Carole was very clear that her mentor had a catalytic impact in helping her develop her activism:

> Ms. Rina. She is the one giving me support if I have plans. So if it's wrong, she tells me that it is wrong. Then she'll give me a better idea on what to do. Because on my own, as I know myself, I can't do stuff like that. Because of the organizers, we started to believe in ourselves.

Dina and Carol's story shows that activism and resistance can flourish even in the context of uneven power geometries. The community workers made a big difference because they had gained the trust of the local women. They lived with them in the same neighbourhood, under the same conditions. They helped the local women with bureaucratic tasks, like filling out government forms and preparing applications for aid. Later, when the grant applications were successful, the community organizers helped women to develop carpentry skills in order to build their new homes. This was an example of grassroots activism in the sense that it was the local women who set the agenda in line with their priorities. Even though this example stood out as exceptional during our fieldwork, it confirmed that people can become empowered to stand up for their rights. We did come across another case of protest during our fieldwork: 'People Surge' involved a series of organized protests and marches albeit with political party support, which is why I do not include that example here (for a discussion of 'People Surge' see Curato, Ong and Longboan, 2016).

Dina and Carol were able to express their concerns publicly in the well-attended community consultation that was organized by the humanitarian agencies in Tacloban. Speaking in front of two hundred people is an impressive feat especially as our local interlocutors often perceived the community consultations as being 'one-way communication'. The two women were also successful in lobbying the government representative for the delivery of new tarpaulin. We can see that their activism was directed both at the government and the aid sector. The reality is that when our interlocutors expressed grievances regarding the recovery process, they typically included the government (national and local) and the aid agencies in their account. This is often because the local government was responsible for organizing the distribution of aid. As we saw in Chapter 3, *barangay* captains were often responsible for drawing up the aid distribution lists. Our interlocutors were aware of the differences between the national government and the humanitarian sector, but their concerns were more global.

Dina and Carol's experience reveals the double role of mobile phone infrastructures. Dina and Carol were able to use the infrastructures and affordances of mobile phones to amplify their voice and challenge their marginalization. These are the same infrastructures that underpin the technocolonial chatbot experiments discussed in Chapter 4. To understand infrastructures and media technologies we need to move beyond the 'either/or' binary thinking, and recognize that technologies can only be understood contextually in relation to specific social orders and social contexts. Digital infrastructures rework the coloniality of power, but they also afford opportunities for marginalized groups to question the conditions of their marginalization and stand up for their rights. Another way this contestation takes place is through digital witnessing and storytelling.

Digital witnessing and storytelling

Meeting with refugee representatives in Athens in the summer of 2016, as part of the 'digital humanitarianism project', I saw photographs that circulated among Viber and WhatsApp groups documenting the conditions in *Idomeni*, a town in northern Greece where refugees remained stuck after the borders closed. Similarly, following the Moria camp fire in Greece in September 2020, videos from refugees' mobile phones served as evidence to document what happened that night. Devices such as mobile phones have proliferated the opportunities of bearing witness.[6] Marginalized groups, who disproportionately experience aggression and injustice, use their phone cameras to document acts of violence. The most emblematic example of mobile-phone witnessing was when Darnella Frazier, a teenager from Minneapolis, filmed the brutal murder of George Floyd by a police officer in May 2020. The video, which went viral on Facebook, sparked the 2020 Black Lives Matter protests in the US and around the world.

The above examples can be understood as acts of 'connective witnessing', a participatory form of witnessing (Mortensen, 2015). The infrastructures and affordances of social and mobile media turn testimonials into a collaborative process. The replicability and spreadability of digital content involve multiple actors and not just the original videographer. The role of intermediaries is important here. Refugees and other crisis-affected people often share their testimonies or their digital photographs with human rights organizations. Alternatively, when the photographs circulate on social media, they can be accessed by organizations to document human rights violations.

One example of connective witnessing concerns the fire at the Moria refugee camp, which involved video footage produced by refugees and the work of Forensic Architecture, a research agency based at Goldsmiths, University of London, whose members investigate human rights violations, including violence committed by states, police forces, militaries and corporations. Forensic Architecture uses spatial analysis, open-source data, digital modelling, documentary research, interviews and immersive technologies, and works in partnership with grassroots activists, international NGOs and media organizations to carry out investigations with and on behalf of communities and individuals affected by conflict, police brutality, border regimes and environmental violence (Weizman, 2014).

Shortly after the Moria fire, and even before the Fire Brigade had completed their investigation, the Greek police arrested six young migrants, five of whom were minors, and who were collectively known as the 'Moria 6'. The Moria 6 were accused of starting the fire and five of them were subsequently convicted on the testimony of a single witness, who did not even attend the trial.[7] In advance of the appeal trial, Forensic Architecture (FA) were commissioned by the lawyers for the Moria 6 to investigate how the fire developed. FA members sourced and examined hundreds of videos, images, testimonies and official reports and conducted a spatio-temporal reconstruction of the spread of fire through the camp. Most of the videos came from footage shot by young migrants on their mobile phones. FA used the time stamps in the metadata of that footage, which were then cross-referenced with additional open-source imagery to create a video timeline of the events.[8] The FA analysis challenged the original trial evidence.[9]

The Moria camp fire investigation is a good example of the collaborative nature of digital witnessing, bringing together refugees, human rights activists, lawyers and investigators. On some occasions, when there are no human witnesses, the material culture and built infrastructure bear witness of atrocities: a burnt camp, a demolished building, or satellite images can reveal human rights abuses in conflict situations. Chris Heller's and Lorenzo Pezzani's work is a powerful example of what data and infrastructures – as well as witness statements – can reveal about the necropolitics of border securitization. Their film 'Liquid Traces: the left-to-die-boat' is an act of witnessing criminal negligence (Heller, Pezzani and Situ Studio, 2012). The 'left-to-die boat', involved the journey of 72 sub-Saharan migrants fleeing Tripoli by boat on the morning of 27 March 2011. After travelling about halfway to the Italian island of Lampedusa during their first day at sea, the vessel ran out of fuel and subsequently drifted for fourteen days without food or water until landing back on the Libyan coast. Only nine of the passengers ultimately survived. In interviews following the event, the survivors recounted a series of interactions they had had with others while at sea. This included a military aircraft that flew over them, a distress call they placed via satellite telephone, two encounters with a military helicopter and fishing vessels, and an encounter with a military ship. Moreover, the Italian and Maltese Maritime Rescue Coordination Centres, as well as NATO forces present in the area, were informed of the distress of the boat and of its location. All these actors had the technical and logistical ability to assist the boat. And yet, despite the legal obligation to provide assistance to people in distress at sea, which is enshrined in several international conventions, none of these actors intervened in a way that could have averted the tragic fate of the people on the boat.[10]

The 'left-to-die boat' film draws on spatial data, archival research and interviews. It has traced the movements of the boat as well as all other vessels and agencies that came into contact with the boat. By providing the sheer evidence of negligence and violation of international law, the film demands accountability for these deaths that were allowed to occur despite the heightened surveillance. The 'left-to-die boat' film, which is an exemplar of the work of FA, turns the uses of technologies of surveillance on their head and uses them to witness atrocities and hold those responsible to account.

Collaborative witnessing is a form of turning the tables, of reversing the gaze back to the perpetrators. States use networked technologies for surveillance and policing; but, increasingly, activists in collaboration with refugees, migrants and others use similar technologies and infrastructures to document injustices (Medrado and Rega, 2023; Ricaurte, 2019; Risam, 2023). Paula Ricaurte (2019) describes how citizen initiatives in Mexico use data and data visualizations to document the violence perpetrated against people. For example, the site Feminicidiosmx.crowdmap.com provides a detailed account of all the femicides that have taken place in Mexico since 2016, thus challenging the official narrative that claims the lack of data to evidence violence against women. By mapping femicides, the activists not only make the killings visible, but also commemorate and honour the murdered women (Ricaurte, 2019: 360). The 'Torn Apart / Separados' project developed by Roopika Risam and colleagues used data visualization to map the detention sites where migrant children separated by their parents at the US/Mexico border were kept (Risam, 2023).[11] Rather than relying on the data provided by the state, the team of academics and activists 'put the gaze *on* the state' and revealed the infrastructures of immigrant detention (Risam, 2023: 160).

Digital witnessing practices often combine with social media practices such as 'hashtag activism', a term which refers to online activism organized through the functionality of the hashtag symbol (#), which aggregates relevant social media content (Jackson, Bailey and Welles, 2020). In order to bring visibility to the predicament of the Moria 6, activists used the #Moria6 hashtag often in combination with the hashtags #FreeMoria6, #JusticeforMoria6 and #refugeesGR. The latter hashtag connected the plight of the Moria 6 more broadly to news about refugees in Greece. Such hashtags interrupt dominant anti-immigration discourses and produce alternative narratives, drawing on the experiences of refugees themselves (see Clark, 2016 for similar arguments regarding hashtag feminism). By mobilizing the storytelling capacities of social media (Papacharissi, 2020), hashtag activism heightened the visibility of the Moria 6 case and became an essential component of digital witnessing.

A further significant transformation is that mobile phones and social media infrastructures have made the boundaries of refugee camps more porous. Refugees continue to be stuck in camps for years; but mobile phones and social media infrastructures allow them to share what it's like to live in a camp with the rest of the world. Charlie Hill has written about the rap music of the Karen youth in the Mae La camp at the Thai-Myanmar border. Her participants record their songs in the music studios in the camp and upload the video clips on YouTube. The songs document the injustices that the rappers experience, as well as their hopes and aspirations. In some cases, the songs have reached large audiences with some videos exceeding one million views (Hill, 2022). This example is linked to witnessing insofar as it involves the sharing of what life is like in a camp. But it might be better understood as a form of networked storytelling and activism. As stories from the camp circulate on the infrastructures of social media, camp life becomes a lot less isolated compared to what it used to be before the arrival of mobile phones and social media infrastructures. Drawing on the rich traditions of protest songs, rap becomes a form of 'poetic resistance' (Hill, 2022).

We have so far explored resistance in the form of direct protest and activism, and resistance through the uses of digital infrastructures for digital witnessing and storytelling. In the next section, I will continue to unravel resistance through the everyday acts of refusal, non-compliance and appropriation.

Mundane resistance

Non-participation

The hardest thing to observe – and write about – is absence. Yet, what is absent is as important as what is present, especially when we are studying the 'hidden transcripts', or how resistance takes place below the radar in the micropolitics of everyday life (Scott, 1990). In order to make sense of mundane resistance we must begin by examining what is missing. The reason mundane resistance is imperceptible, is because the small acts of refusal, ignorance or evasion are hard to identify and record.

Statistics and surveys have a participation bias in that they only count those who participate. This is not unique to the humanitarian machine. When public opinion research states that 38 per cent of the population are in favour of X policy, it refers to the 38 per cent of those who responded. This may sound obvious, but it is often not mentioned. In other words, those who refuse to take part in a survey are not represented in the measurement of public opinion. Election results are the same: when we hear that a party received 25 per cent of the vote, this usually means 25 per cent of those who voted. If turnout was 60 per cent, this means that the 40 per cent of eligible voters are not represented in the share of the vote. The point here is that non-participation entails invisibility in the world of statistics and enumeration. This is problematic because the reasons behind non-participation are complex and may entail agency. For example, not turning out to vote can be a form of protest, a tacit criticism of the political system or a dissatisfaction with the candidates (Cammaerts et al., 2014). Non-participation demands our attention.

To make sense of mundane resistance we need to move beyond the prevailing 'participation fetish' and try to make sense of the silence, non-participation and refusal among marginalized people. Mundane resistance has various manifestations as we shall see later in the chapter. But, in the aftermath of Typhoon Haiyan, non-participation was the most common form of contestation when it came to digital feedback platforms. As we have already seen in Chapter 3, digital feedback mechanisms were problematic in numerous ways: they extracted data for processes of audit, they imposed Eurocentric epistemic frameworks and, most importantly, they often failed to close the feedback loop. During our ethnography in Tacloban and Sabay most of our interlocutors did not use the digital hotlines. Only nine out of our 102 interviewees submitted their feedback via the digital channels that had been prioritized by the agencies themselves. Although we cannot generalize from a qualitative sample of one hundred, during our ten-month long ethnography we met informally with hundreds of people. These wider ethnographic encounters validate the finding about the low take-up of the digital feedback channels.

The refusal to engage with digital feedback platforms is in sharp contrast with the resources allocated to 'accountability to affected people' (AAP) initiatives. Recall that fourteen people worked in the AAP team of one of the INGOs based in Sabay and that all INGOS and aid agencies we encountered during our fieldwork had an AAP programme. Posters advertising feedback hotlines could be found everywhere during our fieldwork, especially on the island of Sabay: on public buildings, local transport hubs, in temporary settlement areas and by the beach. Posters were stuck on walls or the trunks of trees, stitched on tents or distributed as leaflets. In Chapter 3 we encountered one accountability officer who had been advised that their first priority after arriving in Tacloban should be to set up a 'feedback hotline'.

Despite this investment, hotlines were underused at best. Most of my interlocutors had no clear understanding about the purpose of the platforms, which further confirms their Eurocentric character. Given that there is no tradition for asking for feedback in the Philippines, the presence of the hotlines seemed strange, to say the least. For others, feedback hotlines seemed irrelevant to people's daily priorities, which focused on survival and making ends meet in extremely challenging circumstances. A few participants pointed out that the hotlines were not really about them, but a bureaucratic process internal to the agencies.

The low take-up of feedback platforms was also mentioned by some of my interlocutors in the 'digital humanitarianism project'. Reflecting on the Nepal earthquake response, my interlocutor from the aid sector remarked that 'it's all about trust, how much people trust you to share information with you. [...] If they don't trust you, they will not participate'.

Similar observations can be made about the response to chatbots or other platforms. Of the thousands of apps created for refugees since 2015, very few are still functioning at the time of writing. In Chapter 4, we examined Refugee Text, an award winning chatbot that was displayed in the Design Museum in London. Yet, despite its international recognition, the chatbot closed not long after it had launched. The websites of several of these innovation projects constitute an interesting archaeology of digital relics that exemplify the disconnect between self-styled 'digital humanitarians' (Meier, 2015) and refugees. An interlocutor from the aid sector noted that pushing new apps onto people was a 'form of paternalism'. Another aid sector participant noted:

> I like to work with things that people are using. Not the other way round. Like I don't want to make something and then push for something to be used. If you want to be humanitarian you want to build trust and work with what people are already using. [...] And that building of trust is so crucial [...] but it's also like yeah, will [a new app] be used? Will it be useful if I engage with this app? That's also trust. And I feel like it's very often driven by the tech side but not by what people are using.

Non-participation says rather a lot about digital innovation in relief operations. It can reveal lack of trust, disdain, or even a complete disconnect between the shiny platforms and people's actual circumstances and needs. The Eurocentric approach in technology design (Chapter 4) can further explain why platforms seem irrelevant, or even weird, to local people. When it comes to feedback platforms or chatbots, non-participation can send a powerful message. If these platforms or applications are not used, the whole project is cast under doubt. What is the validity of a survey, if the response rate is ten per cent? The same cannot be said about biometric infrastructures, however, where non-participation, at least in situations of encampment, is not an option. A more optimistic story comes from Ukraine, where activists and NGOs were able to refuse the collection of biometric data by the UN in 2023.[12]

Hidden transcripts

Another reason our interviewees did not use the feedback apps is because they constituted 'open transcripts' (Scott, 1990): people were aware that the messages could be read by those in hierarchical positions. In the 'Haiyan project' several of our interlocutors expressed anxiety that any negative criticism might lead to the loss of their aid entitlement. Participants were also concerned that expressing critical views could have repercussions for their family members. These fears were expressed in relation to the government, which was heavily involved in the distribution of aid. We observe yet again the close relationship between humanitarian agencies and the national and local government. Relief was often filtered through the local government, which for some participants created a perception of collusion. Although participants did not mention NGOs specifically when expressing fears about being 'struck off lists', the prevailing attitude was that it was best to 'avoid trouble'.[13] Other interlocutors had internalized feelings of powerlessness. Speaking up was pointless, they said, as 'no one will listen to us. We are poor'.

In asymmetrical situations, power dynamics silence dissent. This should not imply that dissent does not exist. In our Typhoon Haiyan fieldwork, we came across very critical views about the response. Many of these comments were made during the research interviews, or during informal chats in the course of people's daily lives. This is a clear example of a 'hidden transcript' where dissent is expressed in the spaces where those in positions of power are not present. People did not hold back in their private conversations. This is where we encountered grievances regarding the selective nature of aid and anger about the slow pace of the recovery. Criticism also addressed corruption – especially the fact that *barangay* captains, who often proposed the distribution lists, prioritized their families and supporters. We also encountered complaints about the decisions regarding the relocation of families from the most affected areas. Fisherfolk, who lived by the coastal areas that were worst hit, were being relocated in rural, landlocked areas north of Tacloban, with no transport infrastructure, raising concerns about people's livelihoods. While the decision about the site of relocation was made by the government, some of the housing was provided by NGOs (Habitat for Humanity among others) – revealing yet again the enmeshed nature of government and non-governmental activities.

All the above reveal the inherently political nature of relief efforts and the blurred boundaries between state and humanitarian responsibility. While most criticism was directed towards the government, humanitarian agencies were also criticized. On the island of Sabay, some participants even confided in us their wish that the aid agencies had never come to their island, as the selective nature of aid distributions only caused resentment and division among neighbours. Recall that Sabay is where we came across the ubiquitous 'thank you shrines' dedicated to aid agencies, which we discussed in Chapter 3. The stark contrast between the public articulation of gratitude, with its colonial underpinnings, and the private expression of damning critique exemplifies the notion of the 'hidden transcripts'.

Criticism of the recovery process was also expressed on social media and Facebook posts in particular. These posts constitute a 'hidden transcript' despite their semi-public nature as the imagined audience is friends and relatives. By adding their personal experiences to Facebook threads about the recovery process, people produced a collective critique that validated each other's predicament. While these posts were not as vocal – and not as many – as the ones we encountered in the private, in-person conversations, the public articulation of these experiences helped create feelings of solidarity and recognition. These observations echo findings by Anwar and Graham (2020b), who analysed WhatsApp groups as a 'hidden transcript' of informal organizing and resistance among gig economy workers in South Africa, Kenya, Uganda, Nigeria and Ghana.

Appropriation as resistance

Another way local people were able to express their agency was by challenging the intended uses of feedback platforms. The fields of media anthropology and the social construction of technology are replete with examples of how people appropriate technology in ways that are not anticipated by its designers (Horst and Miller, 2006). The example of humanitarian radio that opened the chapter exemplifies the point about appropriation as resistance. A similar story from the 'Haiyan project' includes another humanitarian radio station, which hosted a karaoke radio programme once a week. By investing platforms with alternative meanings, people in Tacloban and the surrounding areas were able to contest the epistemic coloniality of feedback mechanisms.

The examination of the appropriation of technologies must take into account the everyday uses of all communications infrastructures such as mobile phones and the broader online environments people inhabit. Understanding what people actually did with communication infrastructures casts light on their priorities during the aftermath of Typhoon Haiyan. Further, making sense of the technologies that actually helped people navigate the post-disaster landscape of Typhoon Haiyan, also provides an indirect commentary on those platforms that were less relevant. Ironically, the dedicated 'humanitarian' platforms appeared to be the least useful for our interlocutors in the Haiyan project. In practice, people used social and mobile media for a variety of reasons, which did not always have an explicit 'recovery' agenda. As always, technologies are invested with meaning from the bottom-up and 'emergency' contexts are no exception. It is patronizing to assume that people in precarity will only use technology for instrumental purposes or for economic benefit. This is one of the critiques of the ICT for international development agenda which prioritizes the informational uses of technology as though people in low income countries are not as interested in sociality and entertainment as their global North peers (Arora, 2019).

Once internet connectivity was restored in Tacloban (about two months after the Typhoon made landfall) large queues were formed outside the few internet cafés that were able to reopen. People were limited to fifteen-minute sessions initially, before being allowed longer slots. With over ten per cent of the population working overseas and a large Philippine diaspora, many families wanted to reconnect with their relatives abroad. The immediate priority was to exchange news, and update relatives about the situation in the Philippines. Having relatives abroad was considered a blessing, as people depended on remittances to rebuild their homes or their businesses. Inevitably, several calls and messages during this period were calls for help to which the migrant relatives felt obliged to respond. Mobilizing transnational networks of kinship and solidarity was the most effective coping mechanism for families whose members had access to such resources. Migration remittances are the second largest source of income for the Philippine economy, which attests to the privatization of welfare and the failure of the Philippine state to provide for its citizens. Of course, migration perpetuates inequalities as only those who are already privileged can afford to undertake the project of migration. As a result, families with overseas relatives tend to be better off and further benefit from remittances, which in turn reproduce inequalities (Madianou and Miller, 2012; Parreñas, 2001).

People had additional motivations when queuing outside internet cafés. Gamers were keen to return to their online gaming communities from which they had been cut off for weeks. Our interlocutors reflected on the pleasure of gaming as well as the importance of community, which they had missed. Participants wished to reconnect with their fellow gamers with whom they had often formed significant bonds. There were many more online and mobile phone practices – I am inevitably selective here. Many of these practices constitute 'tactics' in the sense that they allowed people to navigate the aftermath of the recovery (de Certeau, 1988). There is an instrumental element to these tactics in the sense that they represent coping mechanisms. But we should not lose sight of the affective dimension of keeping in touch with one's family, friends and community. Building bonds and solidarity can be as important as requesting a remittance to rebuild one's home. Some practices may be fundamentally ambivalent but, ultimately, what matters is the expression of agency and the desire for a return to the rhythms of daily life before the storm.

One practice that deserves some discussion was the use of social media for mourning and memorializing the dead.

Mourning and memorialization

With over 6,300 dead, the loss experienced by people in the affected areas and especially the city of Tacloban in the aftermath of Typhoon Haiyan was immense. Many bodies were never recovered, while in some cases all belongings were washed away: clothes, jewellery, photographs, family albums, birth and marriage certificates; in some cases, whole houses disappeared in the five-metre waves that crashed on the city of Tacloban and the surrounding areas. In some of those cases, all that was left was the digital footprint of the deceased and especially their Facebook profiles, with their selfies and recorded interactions. These profiles became a focal point of mourning and enabled Philippine rituals of memorialization to take place.

In the Philippine context death and mourning rituals are public affairs. This is particularly the case in rural and low-income areas where people are likely to die at home and where death rituals are handled by the family and the community (Cannell, 1999). In her ethnography of the Bicol region in the Philippine lowlands, Cannell observes how the funeral procession (*dapit*), which passes through the whole village, signifies the importance of the deceased. The *dapit*, together with other funerary rituals, is not only a form of collective mourning, but also a tangible way of displaying the status of the deceased. The boundaries between the dead and the living are permeable in Philippine culture, and this is evidenced in funerary and memorialization rituals.

The social media profiles of the dead became a focal point for mourning and memorialization. This was accentuated in the absence of funerals when bodies were never recovered. Over 2,200 bodies were buried in mass, unidentified graves such as the one in *Barangay* Basper of Tacloban city. During our fieldwork we observed that the social media profiles of the deceased remained active. People tagged their dead relatives as a way of 'continuing the conversation'. While in countries like the UK, the content permanence of social media sits in sharp contrast with the finality of death – in the Philippines, content permanence was actually welcomed. The longer the list of comments on one's wall, the greater the sense that they were respected, loved and remembered. Just like during in-person funerals the length of the procession (*dapit*) reveals the status of the deceased, in social media environments, the length of comments on the profile wall is also tangible proof of the respect in which they were held. Even two years after the storm, the Facebook profiles remained active with comments.

I include this example in a chapter about mundane resistance because it shows how social media were filtered through local idioms of mourning. In this example, my interlocutors appropriated social media in a way that was compatible with their ideas and values about what constitutes a respectable way of mourning and memorialization. These are the same social media infrastructures that enable the feedback chatbots, which we analysed as imposing a framework of coloniality. The Filipino way of mourning on social media is a form of mundane resistance insofar as it asserts a local set of rituals and way of being in the world. Again, we see the double character of social media: depending on the context, they can be the means to intimately express deep feelings of loss in a culturally specific way – as well as the infrastructure for the hardly used, paternalistic chatbots.

Resistance or reproduction?

The obvious question to ask about mundane resistance is whether it is transformative, or whether it ultimately reproduces, or even reinforces, existing power structures. If outright defiance is rare in asymmetrical settings, isn't mundane resistance a confirmation of the relationships of domination? It is true that several of the examples discussed in this chapter, such as avoiding digital feedback platforms, do not reverse the power relations between affected people, aid agencies and other stakeholders in the humanitarian space.

The 'safety valve' hypothesis contends that small acts of resistance might diffuse discontent and the grievances generated by relations of domination without engendering social change. Rituals such as the carnival function as a release of social tensions in a relatively harmless way, only for everyday life to resume immediately after (Scott, 1990). In liberal democracies, forms of protest (rallies, marches or petitions) offer a legitimate outlet for the expression of discontent. Some have argued that the tolerance of dissent in Western democracies can reinforce dominant power relations (Death, 2010).

What the safety valve argument ignores is that resistance takes place over time and along a continuum that may include revolution at one extreme, and disengagement or foot-dragging at the other. Outright rebellion does not occur in a vacuum. It is preceded by several small acts of contestation before building up gradually to explicit defiance. Revolution is a process, not a single event. Further, the boundaries between rituals and full-scale revolts can be porous. Scott observes that slave rebellions began precisely during seasonal rituals (1990). It is important to remember that the relationship between symbolic enactments of resistance and practical resistance is a dialectic one (Scott, 1990). There are parallels here with the study of social movements. Melucci has observed that social movements oscillate between periods of visibility and latency (1989). While there is no mobilization, during their latent phases social movements mature and emerge stronger in the next burst of visibility (Melucci, 1989: 71). While there is no clearly articulated anti-technocolonialism movement here, it is impossible to predict whether the mundane acts of resistance may eventually contribute to more systematic forms of questioning or contestation.

It is easy to dismiss the appropriation of humanitarian radio for sociality and entertainment as of no consequence. But it would be misguided to do so as it would ignore the agency of the people of Leyte. Even though mundane resistance does not reverse power relations, to ignore it would amount to silencing those affected by Typhoon Haiyan. To solely acknowledge the dominating structure of technocolonialism would magnify the inequities at stake. This may sound paradoxical given all researchers who focus on inequities aim for social justice. But unless we recognize the agency of local communities, we render them voiceless and helpless – and inadvertently commit the same mistake we set out to criticize.

Conclusion

This chapter has outlined the contours of 'mundane resistance'. Recognizing that technocolonialism is inherently contested, I turned to the Black radical tradition and postcolonial theory in order to make sense of resistance in asymmetrical settings.

By mundane resistance I refer to the dissent that takes place below the radar, through the small acts of everyday life such as refusal, non-participation, storytelling and the appropriation of digital technologies in unexpected ways. Although I draw on Patterson's (2022) discussion of 'passive resistance' in the context of slavery, I argue that there is nothing passive in the micropolitics of everyday life. Non-participation and the refusal to use digital infrastructures can be a political statement, with the potential to even undermine the credibility of some techno-colonial projects. Mundane resistance is also evident in the coping tactics of post-disaster recovery and the appropriation of digital infrastructures for solidarity, community and the articulation of a local way of being.

Resistance emerges along a continuum of direct activism on the one end, and mundane resistance on the other. This is why the chapter also traced other moments along the continuum, including activism, digital witnessing and storytelling among refugees.

The chapter also observed that the opportunities for resistance are asymmetrically distributed within the humanitarian space. For example, non-participation is almost impossible in the case of biometric enrolments in refugee camps. While it's straightforward to simply ignore a digital feedback platform, it is not possible to ignore the request to submit your biometric data or to engage with biometric authentification when this is the only way through which aid is made available. When people are given no alternatives, then non-participation is impossible. There are crucial differences between people recovering from climate emergencies, however devastating, and people living in spaces of securitization such as refugee camps. Technocolonialism and resistance are manifested differently depending on the type of humanitarian context.

Even in camps, however, we observe practices like protest music, storytelling and even acts of outright defiance (Hill, 2022). To equate disaster zones or refugee camps with 'states of exception' is to deny people agency and the possibility for social change (Agamben, 1998). The idea of 'exception' also denies the humanity of people in camps or disaster zones. During the Haiyan fieldwork we attended birthday celebrations, anniversaries, family meals, *barangay fiestas* and karaoke competitions. Ordinary life went on and was celebrated even in the face of extreme adversity. Mundane resistance allows us to discern the small acts or tactics, which may not bring about immediate change, but may pave the way for more spectacular manifestations of resistance in the long term.

Recognizing the agency of communities affected by disaster, epidemics or war is not the same as arguing that they are as powerful as the humanitarian machine. Acknowledging mundane and latent resistance does not mean that the harms of structural violence are resolved. Quite the contrary. People's agency does not cancel structural violence – the same way that structural violence does not obliterate human agency. In fact, the two co-exist; they are shaped relationally. We need to move beyond the binary thinking that we must choose between approaches that either favour powerful structures or human agency. It may seem obvious, but (some) academic fields – and popular discourse, in general – seem to forget that we don't actually have to choose. Structure and agency are co-dependent and cannot be understood in isolation. Some will insist that agency which doesn't effect significant social change simply helps to preserve a problematic social order. This is beyond the point. Mundane resistance does not reverse relations of power, but unless we recognize human agency, how can we ever hope that any social and political change will ever take place?

Conclusion

Technocolonialism as Infrastructural Violence

This is a book about the ways that the datafication and digitization of humanitarianism are reworking the coloniality of humanitarianism and technology itself. I have proposed a new term, technocolonialism, to refer to how digital infrastructures and AI together with humanitarian bureaucracy, state power and market forces reinvigorate colonial structures and produce new forms of violence that shape the relations between the global South and the global North.

In earlier chapters we observed that the reworking of colonial relations takes place through the extraction of value from the data of refugees and other crisis-affected people. Extraction includes biometric data, feedback data as well as data from experimentation with untested technologies. Feedback data are extracted to legitimate humanitarian projects by contributing to audit trails that donors demand. Biometric data and data resulting from experimentation are also extracted for profit, or to satisfy the logic of securitization driven by nation-states.

A further way in which colonial relations are reworked is through the epistemic violence involved in the experimentation with untested technologies such as AI and chatbots. Because AI systems and algorithms are trained in the English language, they impose Anglocentric categories on local cultures. 'Surreptitious experimentation' emerged as a particularly problematic form of epistemic violence as it takes place unannounced, with no accountability. Overall, the lack of meaningful consent and the poor data safeguards across a number of digital programmes magnify the harms from extraction and experimentation. The presence of feedback platforms and digital consent legitimate digital interventions through a veneer of accountability and participation. Finally, colonial relations are reworked through structural violence. Because they depend on classifications that are biased in terms of race, gender, ethnicity and ability, digital infrastructures discriminate. By imposing a Eurocentric gaze on othered bodies, infrastructures produce subjectivities. We encountered such harms in the case of biometric identification errors and the exclusions resulting from automated decision making. Instead of ushering in the 'participation revolution' promised by the World Humanitarian Summit, digital technologies and data practices emerge as means of control and containment of emergencies and crisis-affected people.

The increasing digitization and datafication is so pervasive that it is best understood through the notion of infrastructures, an expansive analytical category that refers to 'built networks that facilitate the flow of goods, people or ideas and allow for their exchange over space' (Larkin, 2013: 328). Infrastructure is a suitable analytical lens because it encompasses the systems (such as blockchain, biometric databases, social media networks, cloud computing), technologies (AI, machine learning, biometric technologies, among others), applications (chatbots, hotlines, apps), data (feedback, biometric or personal data and logistics) as well as the associated practices. All these intersect in complex ways and bring together affected people, humanitarians at different levels of the organizational hierarchy, government officials and private-sector representatives. My research involved following the trails of data: from crisis-affected people and refugees, to humanitarian organizations and then to donors. The infrastructural approach illuminated these data trails and the power relationships at stake. It allowed me to include the micro level of practice (people's uses of the various technologies and applications), the institutional level of policy and implementation and the relations between the different stakeholders.

In this chapter, I want to draw out the implications of the digital and computational infrastructuring of the humanitarian space. In so doing, I will tease out some further themes from the earlier analysis. I will focus on three key implications. First of all, we are witnessing a structural transformation of humanitarianism and its relationship with states and the private sector. Second, we are encountering a new form of violence, which I term 'infrastructural violence'. At the same time, and this is the third implication, the digital infrastructures normalize and legitimate this infrastructural violence. These three observations capture the essence of technocolonialism.

The infrastructuring of humanitarian space and the porousness of humanitarianism

By infrastructuring I mean that technological networks underpin the humanitarian space as a whole. Infrastructuring implies that humanitarian systems build on privatized or government networks and systems. For example, 'communication with communities' and 'accountability to affected people' projects increasingly depend on the infrastructures of messaging applications such as WhatsApp which are owned by private corporations. Biometric systems which underpin cash distribution programmes in refugee camps and other humanitarian settings are run by private companies. One such example is IrisGuard, a company involved in border securitization, which runs *Building Blocks*, the largest biometric cash assistance programme for refugees.[1] Infrastructuring means that humanitarian systems become interoperable with other systems owned by governments or private companies.

The affordances of technologies play an important role here. The immutability and reusability of biometric data means that refugee data originally collected for relief assistance, can end up in other systems, such as counter-terrorism lists (Jacobsen, 2022). Given the interoperability of humanitarian systems, and given the reusability, replicability and retrievability of humanitarian data, the boundaries of the humanitarian space become porous. If humanitarian data are reused and repurposed by governments, and if this flow is facilitated by the interoperability of systems and infrastructures, then where does humanitarianism end and where does state power begin? The boundaries between humanitarianism, states and private companies have always been porous. But the sharing of infrastructures suggests a more systematic synergy. This has important implications for the governance of humanitarian organizations and the integrity of their identity given the priority of relief organizations has always been to distance themselves from state power and to appear apolitical. The irony is that humanitarian organizations have turned to computation and AI to bolster their claims to neutrality and objectivity. Yet, it is these very technological systems that *infrastructurally* bring humanitarian organizations closer to state power. Infrastructurally here means that there is a system, an avenue, in place for these synergies. It is a potentiality, opened up by the presence of infrastructure. We have already seen evidence that these synergies take place (through data sharing for instance). Haggerty and Ericson remind us that as infrastructures transcend institutional boundaries, 'systems intended to serve one purpose find other uses' (2000). I am not suggesting that humanitarian organizations share everything all the time with governments, or their commercial partners. But it is important to recognize that there is now a new substratum that facilitates the flow of data. Regulating these shared infrastructures is imperative.

Infrastructure brings together all stakeholders of the humanitarian space into an entity which I call the humanitarian machine. Because of the shareability of data, and the porous boundaries of responsibility, the machine cannot only be synonymous with humanitarian organizations. It involves the state, as was evidenced by the case of 'double registrations' in Kenya (Chapter 5), or it can also involve private companies, as in the example of biometrically enabled cash assistance through *Building Blocks* (Chapters 2 and 4). This transformation has fundamental implications for the nature of humanitarian organizations which have always strived to distance themselves from the state and the market.

Even though the machine involves different actors, from the point of view of a refugee, the encounter is with one machine which has the power to shape their lives. The fact that there are different 'stakeholders' behind the machine operations is irrelevant for someone who is being denied their citizen status. In the 'double registration' case, even though it is the government of Kenya that denies Somali Kenyans their citizenship, the problem started because their biometric data – collected as part of relief efforts, by relief organizations – were subsequently shared with the Kenyan government.

Such blurring of boundaries has implications for accountability. In the labyrinthine system of vendors, partners, donors, managers and others, who is responsible when a machine error happens? Humanitarian organizations have traditionally been guided by the imperative 'do no harm' which means that they should not put the people under their care in harm's way because of their actions. When humanitarian organizations collect foundational data, like biometric data, to provide relief, the 'do no harm' principle applies. This is because foundational data are sensitive and immutable, which means that they can cause a lot of harm to data subjects if they fall into the wrong hands, or if they are misused.

Given refugee data are permanent, immutable and replicable, and the systems are interoperable, then should governments inherit the 'do no harm' imperative? The infrastructuring of the humanitarian space means that all interested parties would need to subscribe to the 'do no harm' imperative for it to be meaningful.

The humanitarian machine depends on complex supply chains, which obliterate responsibility and accountability. This has implications for the moral relationship between humanitarians and affected people. As automated decision making replaces decision making on the ground, the engagement with actual communities and their problems is undermined. This also impacts frontline staff who are in the field because they are motivated by a desire to help. The hierarchical nature of aid organizations limits the degree of autonomy frontline staff can exercise in relief operations. Policies are directed top down and field staff are expected to implement and report. The introduction of automated decision making introduces new hierarchies in humanitarian organizations and further marginalizes those who are in close contact with communities and who often have a good understanding of their problems. It is revealing that in the example of 'automated decision making' for determining eligibility for assistance (Chapter 5), frontline staff were in the dark about how the algorithm works and how it came to select some families over others. Staff had been tasked to implement a programme which they did not understand and, therefore, could not explain to people in their care. Field staff are most typically local staff and so practices of automation and algorithmic decision making can devalue local knowledge and undermine localization efforts.

The infrastructuring of humanitarian operations introduces more distance between affected people and humanitarian organizations. It is revealing that the introduction of digital infrastructures, such as biometrics, is animated by a suspicion towards 'beneficiaries' and a parallel 'faith' in the capacity of technologies to tell the truth, to be accurate, objective, neutral and efficient. While suspicion predates the implementation of biometrics, it matters if a whole system – or an infrastructure – is designed around 'suspicion'. Suspicion by design will only turn any machine biases (AI is known to discriminate according to race, gender and other categories) into evidence that someone is trying to cheat. In other words, if a refugee is denied aid because of an erroneous match, the most likely interpretation will be that the refugee has tried to game the system; the reliability of the algorithms and the artificial neural networks is rarely questioned. Because AI is cast as superior, the power asymmetrics of humanitarianism are magnified.

I have already outlined some of the implications of infrastructuring for the humanitarian sector. There are additional implications for the humanitarian principles of humanity, impartiality, neutrality and independence we discussed in Chapter 1. If the provision of aid is conditional on submitting one's biometric data, can the principles of humanity and impartiality be upheld in the context of biometric registrations? According to the principle of humanity, all people are eligible for aid with no conditions attached. The principle of impartiality mandates that the only criterion for the provision of aid is need. But if aid becomes dependent on giving one's biometric data, we can discern serious implications for the principles of humanity and impartiality.

Similarly, the capitalist nature of humanitarian technologies has implications for the principle of independence. If a platform is for-profit and the company is answerable to its stakeholders, can a humanitarian project claim independence? When donor countries push for biometric enrolments and fund such initiatives, this raises questions about the independence and the neutrality of the intervention.

The humanitarian principles have always been contested. Despite their limitations, the principles have given protection to humanitarian staff working in war zones. However, the current structural transformation of the humanitarian space entails a fundamental questioning of the applicability of these principles. One implication of this book is the need rethink both technology and humanitarianism.

Infrastructural violence

The infrastructuring of humanitarian operations causes harm. Some errors can have devastating consequences, as in the case of the double registration in Kenya. Mistakes in biometric authentification can cause someone not to receive vital assistance. Being erroneously excluded from aid distribution because of algorithmic errors can entail being deprived of the basic means for survival. Because the machine is cast as scientific, and therefore superior, redressing these errors can be very complicated and laborious. A system error can happen easily, but undoing it can be extremely complicated with the burden of proof falling entirely on the individual. Correcting an error is a protracted and complicated process as the Somali Kenyans who were denied citizenship discovered.

Most errors are less dramatic, but they, too, are pernicious. They involve daily humiliations like being given electronic vouchers that do not function in a rural area that has very few ATMs where one can cash them (Chapter 5). They involve not receiving a response to a text-based complaint, or at best only receiving an automated response that one's 'feedback has been received with thanks' (Chapter 3). They involve being forced to interact with a platform that presupposes mobile or internet connectivity, or even more fundamentally, the ability to write (Chapter 3). They involve encountering a chatbot that does not include one's problem in the 'drop down' list (Chapters 3 and 4). They involve having to interact with a platform that does not reflect local values and imposes a Eurocentric idea, be it accountability or electronic money (Chapters 2 and 3). Presenting people with systems that don't make sense in a local culture, or apps that presuppose an alien understanding of mental illness, is a form of epistemic violence and, therefore, coloniality.

In previous chapters I analysed these everyday violations as a form of structural violence (Farmer, 2005). Structural violence refers to the 'social machinery of oppression' which is exerted 'indirectly, yet systematically' on marginalized groups (Farmer, 2004: 307). Violence affects certain groups systematically (because of their race, gender or where they are located in the world) but is experienced indirectly. While victims experience the violent act in a very personal way, it is impossible to identify a single perpetrator. As Gupta puts it, structural violence is like a 'crime without a criminal' (Gupta, 2012: 21). The diffused yet systematic character of structural violence is particularly pertinent for making sense of the negative consequences of humanitarian infrastructures. When someone is excluded from aid distributions due to algorithmic errors it is not easy to identify who is responsible. Is it the designer of the algorithm, or the organization that decided to use algorithmic decision making? Is it the government data which may be incomplete, or is the mistake the result of problematic classifications about what constitutes a household? The list goes on. What matters is that those at the receiving end experience the blow in a very personal and very direct way. The contact with a machine that refuses to authenticate you while you're waiting for essential relief is crushing.

What also matters is that violence is systematic (Farmer, 2004: 307). The distribution of structural violence is not random. Structural violence is the result of processes of marginalization and social stratification. It reflects specific social, political and economic orders, and affects disproportionately those who are already marginalized and oppressed. Structural violence reflects 'pathologies of power' which so often determine 'who will suffer abuse and who will be shielded from harm' (Farmer, 2005: 7). For example, errors in biometric measurements or identification mostly affect 'othered' bodies whether in terms of race, ethnicity, gender, sexuality, age, or disability (Browne, 2015; Benjamin, 2019). The 'pathologies of power' that Farmer refers to have deep historical roots and are reworked through contemporary practices in the present moment. Border regimes and the idea of the camp itself are two symptoms of these colonial pathologies of power, what Stoler calls 'imperial debris' (2016).

I argue that the process of infrastructuring turns structural violence into an even more diffused phenomenon which is encountered in ever more settings. I term this new type of violence 'infrastructural violence'; it shares all the characteristics of structural violence, but it is even more diffused and multiplied. As infrastructures transcend institutional boundaries and data are repurposed, the probabilities of harm are multiplied. As data infrastructures become ubiquitous, the violence becomes ever more present and diffused – almost like a form of ambient violence. What is also diffused to the point of obliteration is any accountability regarding automated decision making. In the book we have seen glimpses of infrastructural violence. Interoperability is still an aspiration, if not a fantasy, for many of the systems I have analysed. But as the pace of infrastructuring accelerates, we are likely to encounter infrastructural violence more often.

My interlocutors in the Haiyan study and the refugees in camps across the world were already marginalized before they came into contact with the humanitarian machine. To use Achille Mbembe's words, they were 'abandoned subjects' (Mbembe, 2017: 3). The underlying issue here is that the world's most marginalized people are treated as a 'superfluous humanity' (Mbembe, 2017: 3). The problem is that refugee camps exist in the first place, or that climate disasters affect disproportionally those who already live in precarity. The encounter with the machine may be secondary, but it is still important to understand what it produces. My argument is that the encounter with the machine reworks, amplifies and even produces new forms of marginalization and violence. This is what I mean by technocolonialism.

Structural violence is structured (directed at specific groups), but also structuring (Farmer, 2005). Technocolonialism produces subjects. The culture of suspicion that underpins biometric authentifications strips people of their 'believability' (Banet-Weiser and Higgins, 2023) and the agency to define themselves. Because of the immutability of systems, errors become permanent, or are very hard to undo. When algorithmic errors occur, it falls on refugees to correct them, which can be an impossible task. Digital identities mean that borders become embodied and enacted in every interaction with the state. Being a refugee also becomes a perpetual identity as the 'double registration' example revealed. Digital technologies mean that encampment is permanent. If someone's data remain in the refugee database, then they will always be traced back to the camp, even if they have physically left, as happened to the Somali Kenyans in the 'double registration' example (Chapter 5).

Digital infrastructures permeate many aspects of everyday life in the refugee camp or the disaster zone, from cash assistance to grocery shopping. As infrastructures become the 'ambient environment of everyday life' (Larkin, 2013: 328) their harms become normalized. The coloniality of power and epistemic violence implied in these interventions produces subjects. This happens through small acts of humiliation: the technology that doesn't work, the e-voucher system that cannot be cashed because there are no designated shops in the immediate area, or the chatbot that doesn't have a pre-scripted answer to one's problem.

Normalizing (infra)structural violence

The digital infrastructures that constitute technocolonialism at once produce infrastructural violence and hide it. This happens through practices like 'accountability to affected people' which is operationalized as digital feedback. The obsession with feedback mechanisms across humanitarian organizations reveals that feedback has an almost ontological significance for the field. The digitization of accountability turns what should be a messy process into a convenient exercise of box ticking. Listening to complaints and accepting responsibility for things that went wrong can be a sobering activity for anyone. Trying to put things right and empower people to become involved in their own recovery is even harder. This is in sharp contrast to how simple and straightforward it is to set up a hotline. Nothing exemplifies the instrumentalization of accountability more than the automation that chatbots entail. By focusing on a menu of pre-determined answers, chatbots produce feedback well before affected communities submit their responses.

The reason feedback is fetishized in the sector is because it legitimates humanitarian projects. Accountability to affected people becomes a participation smokescreen. The implication is that digital feedback hides the power relations at stake. A similar role is performed by informed consent, which should be taken before biometric enrolment – although as we saw in earlier chapters this practice isn't always observed. Informed consent is only meaningful if there is no detriment for non-participation and if there is full understanding of how the data will be used. In Chapter 2 we observed that neither of these principles are upheld. To refuse to give one's biometric data would entail a refusal to be registered as a refugee and to receive aid. This is an impossible choice for people who are stateless and have no alternatives to survive. Having a full understanding of the uses of one's biometric data seems challenging now that we've explored how the infrastructure works. The permanent, immutable and shareable character of biometric data, coupled with interoperable infrastructures and 'for ever' retention practices, make it hard to predict how the data may be used in the future and by whom. In Chapter 2, we observed that technologies like blockchain and biometrically based virtual cash payments were shrouded in opacity. Without viable alternatives and full understanding of the uses of data, consent becomes coercion. Yet, by appearing to be in place, consent policies normalize and legitimate biometric infrastructures.

I should clarify that I am not suggesting that humanitarian workers involved in 'accountability to affected people' intend to harm. Quite the contrary. I have met several accountability officers who are motivated by a genuine desire to help and are the first to acknowledge the limitations of feedback mechanisms. Their power, however, is limited. This is where the argument about the machine becomes relevant. The humanitarian machine acquires its own momentum and proceeds regardless of individual acts of criticism (Chapter 5).

A third way in which infrastructural violence is occluded is through the 'enchantment of technology' (Gell, 1992). Dazzling technologies such as chatbots, virtual reality, virtual cash assistance or cryptocurrency payments enchant with their novelty and distract from the real issues. Here we see the 'poetics of infrastructure' at play where technologies acquire 'fetish-like aspects' (Larkin, 2013: 329) and are celebrated by technology designers, entrepreneurs and humanitarians. The 'enchantment of technology' amplifies the power asymmetries between affected people and the humanitarian machine. The machine acquires a mythical or magical character which heightens existing power inequities. The dazzling character of technologies not only masks power relations, but also depoliticizes humanitarian emergencies.

Not everyone is dazzled, though. In Chapter 6 we observed that these platforms are contested. Crisis-affected people express their resistance through ordinary, mundane acts. Non-participation, which is often misunderstood as apathy, emerges as a deeply political act. The humanitarian machine depends on everyone playing their role. In that sense, non-participation acquires a radical streak. Refusing to engage with a psychotherapy chatbot is to question the validity of the whole project. There are limitations, however, as non-participation is not an option when it comes to foundational systems such as biometrics which determine whether one is a refugee or not and whether one will receive assistance or not. Chapter 6 developed the notion of 'mundane resistance' to capture the small acts of everyday resistance which challenge the epistemic violence of technocolonialism. Other acts of mundane resistance include storytelling and the assertion of local cultural idioms through the appropriation of platforms. By recognizing the agency of crisis affected communities I do not suggest that the onus is on them to reverse the harms of infrastructural violence. The effort should be to abolish the conditions that lead to technocolonial interventions, and humanitarian interventions in the first place – and this necessitates a wider political project. Even though mundane resistance does not redress the harms of the humanitarian machine, it is important to recognize people's capacity to contest power relations and articulate their way of being in the world. Unless we recognize people's agency, we risk reproducing the social orders we have set out to critique and foreclose any hope for social change.

We thus observe the double role of the infrastructuring of humanitarianism. Technologies produce structural violence and at the same time hide it. The enchantment of technology and practices like digital feedback and informed consent normalize and legitimate power inequities. In this context we observe what Stoler calls 'colonial aphasia' (2016), the inability to speak about injustices and harms even though these exist in plain sight. In situations when the language of bureaucracy and technological enchantment dominate, the real issues are obfuscated by the emphasis on benchmarks, bootcamps and pilots. In parallel, we observe that technocolonialism is inherently contested and this is expressed through open protest, digital witnessing as well as mundane resistance.

Technocolonialism and infrastructuring

We discern the constitutive role of infrastructures in reworking the colonial genealogies of humanitarianism and technology and in materializing colonial presence. While neocolonialism is a cognate term, it does not capture analytically the work that infrastructures do in materializing inequities, enacting structural violence and forming subjectivities, while occluding these harms. This is why we need a new term, technocolonialism. Technocolonialism differs from parallel terms such as data colonialism which has either been developed as a metaphor (Thatcher, O'Sullivan and Mahmoudi, 2017), or to refer to a new phase of expanding capitalism and colonialism, driven by datafication and distinct from 'historical colonialism' (Couldry and Mejias, 2019). By contrast, technocolonialism sees continuities between the origins of humanitarianism in the period of imperial expansion of the nineteenth century and its contemporary iterations. Because it recognizes these continuities, technocolonialism focuses on the inequities between the global South and the global North. Fundamentally, technocolonialism results from the intersection between states, capitalism, humanitarian organizations and digital infrastructures themselves. The notion of infrastructuring, which is key for understanding technocolonialism, captures how all actors become enmeshed in the production of infrastructural violence. Ultimately, technocolonialism is a form of violence that harms some of the world's most marginalized people. This violence can take the form of diffused forms of oppression (infrastructural violence); but it can also turn to necropolitics, when algorithms make decisions about who deserves aid, or even physical violence when the state apparatus becomes involved, as happened in the Rohingya example. Once in place, infrastructures acquire agency and do things. Infrastructure offers a semblance of coherence to the field of digital humanitarianism and, crucially, enables the circulation of data which is how extraction and experimentation take place.

The term technocolonialism emerged through the study of humanitarianism – this book can only claim to be about that. But there are other areas where the term might be applicable. Technocolonialism may offer a framework for analysing the field of international development which has many parallels and synergies with the humanitarian sector. Digital technologies are strongly present in international development as evidenced by the acronym ICT4D ('information communication technologies for development') which has become a field of study in its own right. Technocolonialism can also be applied in the analysis of the 'technology/AI/infrastructure for good' phenomena upon which I have already touched in the book, although 'technology for good' extends beyond relief projects. Finally, technocolonialism might offer insights in the study of areas where the South/North inequities come into sharp focus: the technologization of the response to climate change, the study of pollution and waste, the study of supply chains and supply chain capitalism (Tsing, 2009).

Apart from the argument regarding the reworking of the colonial legacies, the book develops a parallel argument about how the process of infrastructuring transforms the humanitarian sector. Infrastructuring imposes a semblance of coherence upon a fragmented sector, governed by competing logics. Infrastructuring makes the boundaries of aid operations more porous and positions humanitarian organizations closer to nation-states and private companies through the increased opportunities for data sharing. Through this analysis, the book contributes to debates about the role of infrastructures in relation to the privatization of the state and the transformation of public goods.

What is to be done?

One critical question remains as we near the end of this book. What is to be done? I have struggled with this question for a long time as I have been thinking through the developments in the field I study. The development of critique and theory are vital tasks, but in an unstable world ravaged by conflict, war and climate change, the book cannot simply end with critique. I take inspiration from Ruha Benjamin, who urges us to 'practise hope' as well as critique about the injustice that surrounds us (2022). While there are no recipes, I offer here some thoughts. I can only conclude this volume by opening up new conversations as my account, like all accounts, is one of 'incompleteness' (Nyamnjoh, 2017). I group these under three headings: reimagining humanitarianism, reimagining infrastructure and reimagining solidarity.

Reimagining humanitarianism?

Humanitarian reform has been a semi-permanent fixture, partly driven by the humanitarian workers own self-reflexivity and critique. It seems a cliché to propose that humanitarianism needs reform. However, unless its structural inequities are addressed there will never be change. This is a very steep task.

Is it even possible to reimagine humanitarianism? Can humanitarianism be otherwise? Every time humanitarianism has faced criticisms, it has reformed by turning to more bureaucracy and more automation. Humanitarianism will not be reformed by improving efficiencies. The goals of efficiency and audit will never address the power inequities which are at the heart of the problem. Reimagining power relations doesn't imply a new localization strategy, but rather a decentring of humanitarianism so that it doesn't flow from the minority to the majority world. It may be impossible to imagine, but unless it happens humanitarianism will continue to be an 'imperial formation' (Stoler, 2016). One response is to take an abolitionist approach, which entails abolishing the structural conditions that lead to the need for humanitarian interventions. While I draw on the work of some authors who adopt this approach, I acknowledge it is less clear what it means in practice. When faced with absolute decimation and tragedy, should there not be an imperative to help?

The ambivalence towards humanitarian assistance is captured by one of the greatest humanitarian crises and tragedies of our times. Since 13 October 2023, following the 7 October 2023 attacks by Hamas in Israel, Gaza has experienced one of the worst humanitarian emergencies on record, with death and destruction taking place at an unparalleled scale and speed.[2] While I type these words, I am struck by images of the WFP trucks full of food, water and medicine queueing at the border between Gaza and Egypt. Only a few metres from the border, millions of people are displaced, many times over, and face bombings, famine and disease with no medical infrastructure in place. They are trapped in a small piece of land, unable to leave. The contrast between the trucks full of aid waiting at the border and the apocalyptic images of unimaginable destruction and suffering within Gaza crystallizes what is at stake. I found myself wishing for the trucks to be allowed into Gaza so that vital medications could reach the wounded, and food and water could reach everyone else.

I also recall reading Fiona Terry's powerful book where she described how the Bosnian Muslims of Srebrenica declared to the UN agencies that they did not need food, but rather arms so that they could protect themselves: 'We have no need of you, we need arms to defend ourselves, your food aid and medicines will only allow us to die in good health' (Terry, 2002: 22). In the 1995 Srebrenica massacre 8,000 Bosnian Muslim men and boys were brutally murdered by Bosnian Serb Army Units. All this happened in the UN declared safe zone.

For conflicts, resolution can only come through political solutions. But some structure is still needed to provide assistance in situations of extreme suffering. One avenue is for humanitarianism to open up to human rights, social justice and decolonial approaches. This is anathema for most humanitarians as it would open the door to politics and would therefore question their neutrality. But as we have repeatedly seen in the book, neutrality is a chimera at best. Neutrality, even when it is seemingly upheld, contains a form of politics as it often strengthens the oppressors and not the victims (Brauman, 2000; de Waal, 1997; Terry, 2002). Human rights approaches foreground justice and therefore are not afraid to engage with politics. They seek to address the roots of oppression – the reasons behind people's suffering and the reasons their suffering is likely to continue unless it is addressed. Human rights and social justice approaches also seek to remove the structural barriers to people exercising their agency in order to address their own problems. This would be a genuine localization effort, which would involve delegating decision-making to local communities and not just tasking them to implement policies that are developed in the minority world. The role of donors in encouraging a decentralized funding structure is key here.

A social justice approach would also escape humanitarianism's preoccupation with the present – 'the emergency' – by adopting a historical lens. The imaginary of 'emergency' performs important work in humanitarian contexts. It occludes the causes of crises, which need to be addressed for any intervention to succeed. Emergency also justifies a whole range of practices, including technological, with little room for scrutiny. While the emergency imaginary is key for how humanitarian organizations define themselves, in practice, only a couple of humanitarian organizations (MSF and ICRC) practise the pure emergency model of humanitarian assistance. Most agencies and NGOs are involved in a mixture of humanitarian assistance and long term 'development' activities. Programmes like *Building Blocks* and other digital identity systems we explored in the book are not emergency programmes. When refugees are stuck in UNHCR's camps for decades, the response programmes cannot simply be framed under the 'emergency imaginary' (Calhoun, 2008). In other words, humanitarianism is already involved in work that is essentially political. Maybe the door to a conversation about a social justice and decolonial approach is not as firmly closed as it appears.

Reimagining infrastructure

Can 'technology for good' ever be for good? Even presumed benign interventions such as drones and satellite monitoring rework inequalities.[3] AI will always depend on classifications that are subjective and which mirror the values of its designers (Gebru, 2020). English language dominates the online world, which is why it is impossible to decolonize AI as long as the algorithms and large language models are trained on huge datasets in the English language.

The narrow definition of good implied in 'technology/infrastructure/AI for good' is inherently problematic. The current dominant understanding of 'technology for good' equates good with the UN sustainable development goals. This will always be paternalistic and Eurocentric. 'Good' carries moral weight and confuses the debate by turning what is a question of power (who defines good, and for whom) into a question of ethics. This was evident in the roll out of biometric digital identity systems which are championed as 'for good' solutions but are actually animated by suspicion towards refugees. In asymmetrical settings such as humanitarianism, unless design is decentred, it will only amplify the power inequities.

The notion of 'good' with its ethical connotations is inadequate for reimagining technology or infrastructure. Rather than 'for good', one proposal is to design for the 'pluriverse', 'a world in which many worlds fit' (Escobar, 2018: 16). The term pluriverse evokes a multicentred world rather than one which replicates the imperial faultlines of North–South, or East–West. Drawing on decolonial theory and its praxis ethos, Escobar argues for the development of 'autonomous design' that foregrounds collaborative and place-based approaches. Autonomous design is not for profit and eschews the universalizing paradigm of modernization that has plagued international development. Design for the pluriverse takes a social justice and 'bottom up' approach to the design of infrastructures and technologies.

There are similarities with the 'design justice' approach outlined by Sasha Costanza-Chock (2020). Design justice is a collaborative approach that rethinks design drawing on social justice principles. Design justice 'centres people who are normally marginalized by design' and 'uses collaborative creative processes to address the deepest challenges communities face' (Costanza-Chock, 2020: 6). One of the aims here is to address the inequalities reproduced by mainstream design. Design in this context is not necessarily about digital platforms or coding; it can involve reimagining infrastructures more broadly. One example might be the design of public infrastructures of care in post-disaster settings.[4] Digital infrastructure projects could include the development of platform cooperatives, or what Nathan Schneider calls 'governable stacks', which are systems that afford self-governance to all their members (Schneider, 2022). Such an example is 'May First Movement Technology', a non-profit cooperative that 'engages in building movements by advancing the strategic use of collective control for local struggles' and offers autonomous technology infrastructure (from website hosting and email to virtual meetings and chat) to all its members.[5] Also relevant here is work on feminist AI as a form of 'resistance to large-scale hegemonic AI' (Toupin, 2024).

Central to all this is the development of a 'critical pedagogy' and 'critical consciousness' that empowers people to become involved in the processes of reimagining (Freire, 1970). Not only is critical consciousness vital for collaborative design, it is also important for the key task of scrutinizing the existing systems. Interrogating how systems work, which classifications they depend on, who has access to one's data and with what consequences, is an important first step before reimagining alternatives. Campaigning for the regulation and transparency of public–private partnerships, and for the reversal of humanitarian marketization would be the logical next steps.

Although questions remain about whether AI systems can ever serve a social justice and decolonial agenda – given their systemic Anglocentric formation as algorithms are trained in the English language – the pluriverse and design justice approaches offer a promising start for imagining infrastructures otherwise.

Reimagining solidarities: Organize!

The discerning reader will recognize the phrase 'what is to be done' as the title of Ruth Wilson Gilmore's 2010 essay. In her address, on which the essay is based, Gilmore asked this question repeatedly and the answer came back like a chorus: organize! Structural violence affects categories of people, but it is experienced individually. The onus is on the individual to seek redress, which is unrealistic. By seeking others, those affected can find validation and can combine forces that seek reparation and redress. Even though solidarity will not resolve the grievances of colonialism, collective action is the only way to begin addressing the violence of colonial structures. The most optimistic moments in this book have come from the examples of resistance and solidarity building: the activists in the 'tent cities' of Tacloban where they mentored women to stand up for their rights (Chapter 6) and the collaborative witnessing between migrants, researchers and activists as in the case of the Moria 6 (Chapter 6). There are different levels of organization here. There is solidarity-building among refugees and crisis-affected people. Additionally, there are opportunities for activism between affected people and human rights activists. We need new forms of 'political care', which Miriam Ticktin defines as 'new ways of being together at a global scale, grounded on participation and labour, duty and obligation and shared common resources' (2016: 267).

Several such spaces have emerged in recent years, especially around migration and refugee rights. Of particular interest to the concerns explored in this book are initiatives that focus on technology. One example is 'Homo Digitalis', a Greek NGO which focuses on the protection of digital rights, whose members uncovered the significant shortcomings of the security and surveillance systems deployed in refugee camps in Greece. The systems in question, both of which are funded by the EU, involve 'Centaur', an automated security system that relies on algorithms and hardware, including cameras, drones and sensors, and 'Hyperion', a biometric technology system. In April 2024, the Greek Data Protection Authority found that both systems had violated several provisions of the General Data Protection Regulation (GDPR) and issued a fine to the country's Ministry of Migration, although it is unclear what that means for the future use of the systems.[6] The following example illustrates how similar activism can be successful in reversing some of the most problematic policies of technocolonialism.

During the war in Ukraine, a group of INGO staff campaigned to refuse to collect the biometric data as part of their response.[7] The argument was twofold: the Ukrainian government insisted that it wanted its law enforced so that its citizens' data would not be collected by international organizations. Civil society groups within Ukraine also expressed strong views against the use of biometrics. INGO staff added a separate line of argument regarding the applicability of the European Union's GDPR legal framework for Ukrainian refugees in neighbouring Poland and Romania. GDPR has special provisions for sensitive foundational data such as biometrics that are not met by the safeguarding criteria of UN data policies. The campaign was successful and biometric registrations stopped in July 2023. While this is an encouraging development, it also confirms that GDPR is a luxury that does not apply to the majority world. The Ukrainian case crystallizes what most data activists and critical data scholars already knew, which is that UN data practices would not pass the GDPR test. There is safeguarding and legal protection for people in the global North, but not for people in the south. Yet the Ukraine example sets a precedent on how activists and NGO workers can campaign to refuse biometric technologies – or any other technologies – that constitute a violation of human rights. This case also demonstrates that there are ways to effectively distribute aid at scale without biometrics. The Ukraine example reminds us that techno-colonialism is not a given and that its trajectory can be stopped.

Technocolonialism is structural. Its harms will not be reversed if we simply tweak the machine or improve the algorithms of automated decision making. It is only through political and collective action that we can address the structural violence that the machine helps to produce – and imagine a hopeful future.

A Note on Research Methods

The book draws on almost ten years of research on the uses of digital innovation and data practices in humanitarian operations. During this period, I led two projects and conducted additional desk research. In total, the book draws on 185 semi-structured interviews, multi-sited participant observation, digital ethnography and additional desk research. In this note I elaborate on the research contexts, methodology and the data collected. Because of differences in research design, I discuss the two projects in turn. All book chapters draw on the research for the 'digital humanitarianism project', conducted between 2016 and 2021. The 'Haiyan study' mostly informs two Chapters (3 and 6), but because of its formative role in terms of how it helped shape my thinking in this area, I will begin with it.

The Humanitarian Technologies Project (The 'Haiyan study')

Between 2014 and 2015, together with Jonathan Ong, Liezel Longboan, Nicole Curato and Jayeel Cornelio, we conducted an ethnography of the aftermath of Super Typhoon Haiyan (locally known as Yolanda), which hit the Philippine archipelago in November 2013 and remains the strongest storm to make landfall. Our ethnography followed the experiences of affected communities and their communication practices over the course of ten months. Apart from local residents displaced or otherwise affected by the disaster, we also interviewed humanitarian officers, government officials, volunteers and activists.

This was a multi-sited project focusing on the city of Tacloban and the surrounding areas and the island of Sabay (anonymized for reasons I explain below). Tacloban is the capital city of the island of Leyte and the largest city in the Eastern Visayas. Sabay is a remote island in the Visayas archipelago, off the coast of Cebu, which depends largely on fishing, although tourism is a growing sector. Tacloban was at the epicentre of the relief efforts having been one of the worst affected areas with a high number of casualties. Storm surge waves submerged most parts of the city with whole neighbourhoods completely devastated. Sabay reported severe material damage, but few casualties, as it was not affected by the storm surge. We arrived in Tacloban in April 2014, five months after the storm and first visited Sabay in June 2014. Team members subsequently moved between the two sites.

During our ten-month ethnography we conducted participant observation and interviews with 102 participants affected by the Typhoon. In order to get a sense of the temporality of the recovery and the associated communication practices we interviewed several of our interlocutors more than once and met with them regularly as part of the wider ethnography. Ethnography is recognized as an appropriate method for research with disaster-affected people (Adams, 2013). The long-term immersion in the everyday life of local communities allows for relationships of trust and understanding to develop. During our fieldwork, we spent time with our key participants and their extended families: we shared meals, celebrated anniversaries, attended local festivals as well as community consultations. We met informally with several other people as part of the ethnography and kept detailed notes from all these encounters which have informed the analysis. We supplemented participant observation with online ethnography in order to make sense of our interlocutors' online practices. As we were able to follow developments for a period of over ten months we were able to develop a sustained account of our participants' experiences and struggles over the course of a year.

We met our participants through our ethnography of selected *barangays* (neighbourhoods) in the two fieldsites. Participants came from a broad range of backgrounds, ages and socioeconomic classes: our sample included 55 women and 46 men; 63 of our participants were very low, or low income, while 38 were middle class. All our participants were affected by the Typhoon, but the majority of low-income participants lived in temporary accommodation even one year following the storm.

The Haiyan study included additional interviews with 38 experts: humanitarian officers, representatives from government departments, local civil society groups, media organizations and private companies. Our sample consisted of large international humanitarian organizations as well as smaller NGOs. We included officers at different levels within the organizational hierarchy (frontline staff, accountability officers, and representatives from the national offices or headquarters). As part of our ethnography we visited agency offices and attended consultations. Notes from these encounters have also informed the analysis.

Interviews lasted ninety minutes on average and were recorded, transcribed, anonymized, coded, and analysed thematically. Because of the sensitive nature of the research, we anonymized all interview data including expert interviews. Expert data are also sensitive, and because we wanted our expert interlocutors to feel comfortable enough to speak freely and frankly about their involvement in the relief and recovery efforts. We also took the decision to anonymize the location of our second fieldsite: Sabay is a pseudonym. This decision was made in order further to protect the anonymity of our participants. Because of Sabay's small size, our research ethics review concluded that our interlocutors could still be identified even if their names and identifying variables were changed. We did not anonymize the city of Tacloban. Because of its size as a regional capital and urban administrative centre, the risk of identifying anonymized participants was very low. The severity of destruction experienced by the people of Tacloban turned the city into a symbol of the Haiyan disaster as a whole. This is an additional reason why it would have been impossible to anonymize the city of Tacloban.

The Humanitarian Technologies Project ES/M001288/1 was funded by the UK's Economic and Social Research Council (ESRC).

The 'digital humanitarianism project'

In 2016 I broadened my focus and began exploring the role of digital innovation and data in the humanitarian sector. As another humanitarian crisis unfolded across the Middle East and Europe, I witnessed the explosion of digital innovation 'for good'. I was keen to further explore the role of technology in the sector that was rapidly including AI powered chatbots, automated decision-making and biometrics.

Between 2016 and 2021, I conducted 45 interviews with seven groups of stakeholders: humanitarian officers, donors, entrepreneurs and business representatives, digital developers, consultants, government representatives, and volunteers. In addition, I conducted participant observation in spaces of innovation, such as hackathons, as well as digital ethnography.

The main research method in the 'digital humanitarianism project' was semi-structured interviews. Interviews gave me insight on the perspective of those involved in aspects of digital innovation in humanitarian operations. I made an effort to speak with people at different levels within the various organizational hierarchies. Although I aimed to include interviews with people from different sectors, in practice these distinctions are not entirely clear cut. Several of my participants wore several hats, or had career changes from the INGO to the government or private sector or vice versa. Most of my participants came from humanitarian organizations, whether intergovernmental agencies (such as UN agencies)or international non-governmental organizations. I am still struck by the generosity of humanitarian officers and their willingness to speak with me.

Interviews were conducted between July 2016 and July 2021 and took place online (via video platforms), or in person (in London, Cambridge, Athens, New York and Washington, DC). My interlocutors were scattered across the world, from the Caribbean to the Middle East, and from Central Africa to Europe and North America, which is why online interviews were necessary. The Covid-19 pandemic, and the social distancing measures it mandated, meant that further interviews had to take place via videoconferencing platforms. Interviews lasted sixty to ninety minutes on average and were recorded, transcribed and analysed thematically. Three interviews were not recorded at the request of participants. In those cases, I kept detailed notes. All interviews were anonymized and are presented here in an aggregate form in order to protect the anonymity of expert interviewees. For example, when I refer to aid sector participants this may include humanitarian workers or consultants. The article also draws on fieldnotes I took during participant observation at twelve industry events, meet-ups and hackathons.

The project also included digital ethnography (Pink et al., 2016). I consulted hundreds of blogs, websites, podcasts, policy documents, social media content and videos of talks, webinars and event recordings. In other words, I 'followed the thing' (Marcus, 1995) across online and offline spaces. I continued to follow the phenomenon of digital humanitarianism through extensive desk research until the final phases of writing the book in the summer of 2023. Digital humanitarianism is by definition a multi-sited phenomenon, and digital ethnography was invaluable in my effort to follow the trails of data and innovation globally. Finally, as part of the digital ethnography, I took every opportunity to interact with humanitarian platforms: from dashboards such as WFP's 'HungerMap Live' to dedicated refugee apps and humanitarian chatbots. I kept records of these interactions and used my experience as an entry point for making sense of the workings of these platforms. In Chapter 4, for example, I discuss how my interactions with chatbots made me realize how the AI programme disciplined me into 'speaking' in a particular 'chatbot-like' way in order to become legible to the chatbot (mostly unsuccessfully). Such autoethnographic insights were invaluable for making sense of developments in the field (see Barassi, 2021 for a similar use of autoethnography). The inability to travel during the Covid-19 pandemic further necessitated this heterogeneous approach to ethnographic data collection which echoes what Seaver (2017) calls 'scavenging ethnography'.

Anonymization of my interlocutors and their organizations was key to my approach. I wanted to make sure that my interlocutors felt comfortable to speak freely about their experiences and thoughts regarding developments in the field. This approach is common in research with experts (Thompson, 2021). Even though classified as experts, I am aware that some of my participants' work in conditions of precarity and I have a duty of care towards them. For this reason, I do not specify the affiliation of my interviewees, nor do I refer to their exact role. By presenting interviewees in aggregate form (e.g., 'a participant from the humanitarian sector'), I add another layer of anonymization.

When referring to data collected as part of the digital ethnography, assuming that the material is already publicly available online, then I refer to the name of the relevant organization and the particular aid projects. This is because the material already exists online and can be easily traced by anyone.

In combination, the two projects gave me a broad overview of digital and computational developments in the humanitarian sector. Combining multiple sites and crises, and bringing together different levels of analysis (from the micropolitics of everyday life, to the perspectives of frontline workers, designers, humanitarian organizations and governments), allowed me to see the development of infrastructures, which by definition transcend institutions and specific locales. My iterative ethnographic approach evolved and expanded as I started to follow the trails of data and innovation. This was necessary in order to be able to map the power relations within the field of study.

Including the lived experience of my participants from the 'Haiyan study' was of paramount importance. The perspective of people affected by disaster, conflict or other emergencies is often left out from debates. There are similar patterns in critical data studies. As the pendulum has swung (again) towards models of data and technological power, the voices of data subjects are not heard as often as they might be (but see Kennedy, 2018). It mattered to me that people affected by crises are not spoken about, but are allowed to articulate their experiences in their own words. I hope that I have managed to convey some of their experiences and that they will be able to see themselves on these pages.

I want to end with a reflection on my own role in the research process as I recognize that all knowledge production is situated, especially given the participatory nature of ethnography, where my experience has been used to produce knowledge. I came to this topic following a long-standing interest in the Philippines, where I have been conducting research since 2007. I came to this research with my own multiple identities as a woman, a mother, a migrant and an academic. I was born and raised in Greece but have lived in the UK for most of my adult life. I have often thought that my long-standing interest in the Philippines might be explained because it resonates with my own experiences. Greece and the Philippines are two very different countries, but there are similarities. Some might point out that, unlike Greece, the Philippines is a postcolonial state. Yet Greece maintains an ambivalent position: it animates an imaginary of the 'cradle of Western civilization' and an orientalist imaginary of a country which emerged after 400 years as part of the Ottoman empire. Greece's independence was underwritten by three guarantor powers (Britain, France, Russia) which openly interfered with Greek politics. The fact that Greece had an imposed foreign king, a history of dept – and defaults – that tied the country's fate to foreign creditors and a brutal civil war (1946–9) is testament to foreign interference and what Herzfeld 'cryptocolonialism' – the disguised forms of colonialism experienced by nations that have been controlled by 'Western powers' but not directly occupied. (Herzfeld, 2002: 923). More recently, during the debt crisis of the 2010s, the draconian austerity imposed by the 'institutions' resurfaced long-standing debates about the status of Greece as either Oriental or European (Mezzadra and Neilson, 2019; Mignolo and Walsh, 2018). Cryptocolonialism points to the existence of 'new subalternities and new complexities of power' which challenge the neat split between colonizers and colonized (Herzfeld, 2002: 923). All this is to say that through my intersecting identities, I relate to the affect of 'othering' through lived experience and not in an abstract way.

At the same time, I fully acknowledge my very privileged position as a university professor based in London. My professional status has opened doors during the research process. I publish articles and travel to international conferences to share my research findings. I have spent months in the field. When I'm there I aim to be part of my participants' lives: we share meals, cook together, we even sing karaoke (at least I try). But there is a fundamental difference in that at the end of my fieldwork, I was able to leave, which was not an option open to my interlocutors, many of whom are still in temporary accommodation or stuck in refugee settlements. This reflexive note is not a solution to the fundamental inequity of the research process (Lehuedé, 2024). The ethnography, with its participatory nature, is my only tool to develop research that is dialogical, empathetic and aspiring to social change. Still, it matters to situate my own participation in the process of knowledge production.

Method is always about much more than the techniques and tools that we use as researchers. It is the lens through which we see the world, the decisions we make about where to look and what to see. Writing the book has been a process of unlearning and realizing some of my own blind spots. I'm sure more remain, for this work, like all projects is characterized by 'incompleteness' (Nyamnjoh, 2017).

Acknowledgements

This book has been a long journey. Some of the people I would like to thank the most cannot be named here because of the promise of anonymity that is essential for ethnographic work. I owe the greatest debt to my interlocutors in the Philippines and, in particular, my friends in Tacloban and Sabay. I cannot thank you enough for your time and humanity, even when you were facing the most urgent demands of rebuilding your lives in the wake of Yolanda (Typhoon Haiyan).

I am also deeply indebted to the humanitarian workers and other officials who generously gave me their time – often more than once – and shared their insights. I include here participants in the Philippines but also in the later 'digital humanitarianism project', on which the book is primarily based. I have learned more from you about humanitarian practices and digital innovation than I would have ever learned by reading articles and reports. I would not have been able to write this book without your input. As always, I take full responsibility for the contents of the book.

I am very grateful to my collaborators from the 'Humanitarian technologies project' which informs two chapters of the book: Jonathan Ong, Liezel Longboan, Jayeel Cornelio and Nicole Curato. Our fieldwork was one of the most illuminating, but also challenging experiences and I would like to thank you for sharing that with me. Special thanks to Jonathan for his insights, creativity and resourcefulness; and Liezel for her continued presence in Tacloban and Sabay and dedication to the project even in moments of adversity. The Humanitarian Technologies project was funded by the ESRC (award: ES/M001288/1).

Between 2017 and 2022 I was fortunate to be part of a generous and generative research community. The Public Religion and Public Scholarship in the Digital Age Project, funded by the Luce Foundation, and hosted by the University of Colorado at Boulder, was a refuge in a time of multiplying crises. I thank Sarah Banet-Weiser, Anthea

Butler, Nabil Echchaibi, Christopher Helland, Stewart Hoover, Marwan Kraidy, Peter Manseau, Sarah McFarland-Taylor, Nathan Schneider, Jenna Supp-Montgomerie and Debbie Whitehead for their intellectual companionship and friendship. The book owes a lot to our conversations.

The research and writing for this book have coincided with my time at Goldsmiths, University of London, which has also funded the 'digital humanitarianism project' through its research themes fund. Most importantly, the Department of Media, Communications and Cultural Studies at Goldsmiths has provided the intellectual freedom and inspiration for my work. Our department is a unique place in its combination of commitment to social change and collegiality. I am grateful to all my colleagues for always inspiring me with their amazing work and for their fierce solidarity, even in the most turbulent times. Des Freedman deserves a special mention for his incredible kindness and courageous leadership as head of department under the toughest of circumstances. I would also like to thank our current and previous heads of department for nurturing the space where ideas and critical thinking are foregrounded: Daisy Asquith, Lisa Blackman, Sean Cubitt, Aeron Davis, Natalie Fenton, Becky Gardiner, Julian Henriques, Damian Owen-Board, Nate Tkacz and Joanna Zylinska. Natalie Fenton has been a wonderful mentor and friend who I can always trust to see through what really matters. Special thanks to Clea Bourne and Liz Moor for sharing our writing journeys; Matthew Fuller for our work at the Digital Cultures Unit; Sue Clayton, Mark Johnson and Michaela Benson for pushing the migration agenda across the College; and Gholam Khiabany for being the soul of the department.

My thinking has been shaped through conversations with my PhD students and postdoctoral fellows, past and present. Thank you for one of the most rewarding aspects of academic work. During the period of working on this book this group included: Silvia Carrasco, Hong Chen, Charlie Hill, Amanda Hope Macari, Callum Morrissey, Andreas Schellewald, Grace Tillyard and Sanja Vico.

The idea for this book was first developed at my keynote talk at the Heyman Centre, Columbia University in April 2018. I would like to thank Sandra Ponzanesi for the generous invitation speak at the 'Migration and Mobility at a Digital Age' conference. A revised version of the talk was then published in *Social Media & Society* in

2019. The book grew out of that article, but the argument has been radically reworked and expanded. Over the years I have been invited to give keynote lectures from this research at various conferences. I'm very grateful to the organizers of the following events for hosting me and to the participants for the lively conversations and feedback: Association of Internet Researchers annual conference in Dublin; Centre for Global Media and Digital Communication Annual talk, SOAS; Data Justice Conference, University of Cardiff; Digital Diasporas conference, School of Advanced Study, University of London; Danish Institute of International Studies, Copenhagen; Gender, Migration and Digital Networks Conference, National University of Singapore. Additionally, I'm indebted to colleagues who discussed this work with me in seminars and workshops at the Universities of Ateneo de Manila, Athens, Bergen, Cambridge, Edinburgh, Fudan, Groningen, Malmo, Lund, Sheffield and the University of the Philippines-Diliman and further afield, as well as various online workshops that took place during the Covid-19 pandemic. One highlight was the Race, Technology and Borders workshop led by Tendayi Achiume at UCLA in June 2020, which provided fertile discussions on the racialization of digital bordering and securitization.

During November 2023 I was visiting Professor at the University of Philippines. It was so important for me to be able to return to the Philippines, reconnect with my interlocutors, and re-immerse myself into everyday life in Manila. I am grateful to Fernando Paragas, Randy Solis, Junel Labor and Shine Rapanot for the generous invitation and for being the most convivial hosts. In Manila I can always count on Jason Cabañes (now our colleague at Goldsmiths), Cheryll Soriano, Anjo Lorenzana and Raul Pertierra to make my visits exciting and insightful. I learn so much from our conversations.

I am forever indebted to Lilie Chouliaraki, Radha Hegde, Zizi Papacharissi, Eugenia Siapera and Cara Wallis who have followed the writing of this book through its various stages and have offered invaluable feedback and support.

The book has been enriched through my conversations with Nancy Baym, Jack Bratich, Benedetta Brevini, John Bryant, Elinor Carmi, Nick Couldry, Lina Dencik, Stephanie Diepeveen, Larissa Fast, Megan Finn, Michael Guggenheim, Sofie Elbæk Henriksen, Lilly Irani, Katja Lindskov Jacobsen, Helen Kennedy, Aphra Kerr, Nauja Kleist, Monika

Krause, Mark Latonero, Tracy Lauriault, Sebastiàn Lehuedé, Koen Leurs, Carleen Maitland, Annette Markham, Dina Matar, Ella McPherson, Andrea Medrado, Ulises Mejias, Philippa Metcalfe, Daniel Miller, Kate Nash, Serena Natile, Kaarina Nikunen, Shani Orgad, Allison Powell, Paula Ricaurte, Lisa Ann Richey, Roopika Risam, Adrienne Russell, Lynn Schofield Clark, Shela Sheikh, Darryl Stellmach, Linnet Taylor, Sophie Toupin, Emiliano Treré, Karin Wahl-Jorgensen, Saskia Witteborn, Kate Wright and many more. I thank them all as well as the anonymous reviewers for their detailed and thoughtful comments, which helped me improve the manuscript.

I am enormously grateful to Zach Blas for allowing me to use 'Face Cages' for the book cover. 'Face Cages' visualizes the biometric diagram and turns it into a cage which is then violently imposed upon the human face. When I first saw 'Face Cages', I immediately thought that it encapsulated what I was trying to convey through the notion of technocolonialism. I am delighted to be able to use the artwork as the book's cover.

At Polity, I would like to thank my editor, Mary Savigar and Stephanie Homer for the care they have shown and for their great patience waiting for the final version of the manuscript. Many thanks to Susan Beer for careful copy-editing and to Evie Deavall for ensuring a smooth production process.

I could not have finished this book without the love and support of my friends and family in Athens and Cambridge. You know who you are and how much happiness you bring to my life. Special mention to my brother George for reminding me I am more than my work. My parents, Michael and Dimitra, have been my greatest supporters. I will always cherish our summer of writing in Antiparos, which was key for getting the final draft into shape. I hope you understand why I had to travel far to write this book – and why it took me so long to write it.

To John and Alex: you give me hope that a better future is possible. Thank you for making the time for me to finish the book and for understanding why it matters. I am now ready for our next adventure.

Notes

Introduction

1 UNHCR 2022 figures at a glance: https://www.unhcr.org/uk/about-unhcr/who-we-are/figures-glance
2 Global Humanitarian Overview 2023: https://reliefweb.int/report/world/global-humanitarian-overview-2023-enaresfr
3 The UNICEF cryptofund: https://www.unicef.org/innovation/stories/unicef-cryptofund
4 UNHCR refugees vulnerability study: https://reliefweb.int/report/kenya/refugees-vulnerability-study-kakuma-kenya
5 UNHCR Project Jetson: https://jetson.unhcr.org/
6 The unofficial estimate is higher, with suspicion expressed locally that the official death toll was deliberately underestimated (Curato, 2019).
7 https://reliefweb.int/report/philippines/joint-statement-humanitarian-country-team-10th-year-anniversary-super-typhoon-haiyan-philippines
8 For a discussion of Skype in the asylum pre-registration process see Aradau (2022) and Metcalfe (2022).
9 I here refer to European colonialism between the early sixteenth century and the 1960s when Western European countries conquered and dominated large parts of Africa, Asia and the Americas. While many of the colonized countries achieved independence by the 1960s, many indigenous people still live in settler-colonial states. Although there are several definitions and types of colonialism a common feature is the understanding of colonialism as the annexation of land and the subjugation of people for the economic benefit of the colonizers. Race played a pivotal role in justifying why certain people did not deserve to be free, therefore rationalising the European colonial project.
10 See Kundnani (2023) for an excellent synthesis of racial capitalism, colonialism and imperialism.
11 I focus primarily on these because they are emblematic of the uses of digital innovation and computation in the sector but the chapters include further examples of innovation such as virtual reality (VR) among others.
12 https://aiforgood.itu.int/
13 The book cover includes the face of Elle Mehrmand under a three-dimensional metal cage, evoking a material resonance with handcuffs, prison bars and

torture devices used during slavery in the United States: https://zachblas.info/works/face-cages/

14 ALNAP, 2022.

1 *The Logics of Digital Humanitarianism*

1 For a comprehensive history of humanitarianism see Barnett (2011).

2 Pictet identified seven principles. Not mentioned here are 'voluntary service, unity and universality', which were operational principles with reference to the ICRC, although they, too, have been influential. For discussion see: https://blogs.icrc.org/cross-files/the-fundamental-principles-of-the-international-red-cross-and-red-crescent-movement/

3 UNOCHA Sudan: https://www.unocha.org/sudan

4 UNOCHA Sudan: https://www.unocha.org/sudan and *Guardian*, 'Inside the Darfur Camp where one child dies every two hours': https://www.theguardian.com/global-development/2024/feb/21/darfur-sudan-zamzam-camp-child-dies-every-two-hours

5 UNOCHA Sudan: https://www.unocha.org/sudan

6 UNOCHA Occupied Palestinian Territory: https://www.unocha.org/occupied-palestinian-territory

7 https://www.gov.uk/government/news/prime-minister-announces-merger-of-department-for-international-development-and-foreign-office

8 ALNAP, State of the Humanitarian System, 2017: https://sohs.alnap.org/blogs/data-story-whats-the-shape-and-size-of-the-humanitarian-system

9 https://executiveboard.wfp.org/document_download/WFP-0000148942?_ga=2.200223918.1048711889.1699192836-1696124980.1692017183

10 https://www.wfp.org/stories/wfp-glance

11 For a discussion of the Rwanda civil war and genocide and the role of the UN and relief agencies see Barnett (2011: 180–5); Rieff (2002: 155–96) and Terry (2002).

12 The 'Grand Bargain': https://interagencystandingcommittee.org/node/40190

13 Ibid.

14 Another relevant figure is that the global Official Development Assistance (ODA) budget, which was US $204 billion in 2022, a huge expansion of over US $70 billion within ten years (the ODA budget in 2013 was US $134 billion). The ODA figure includes the budget for international development projects but, given the porous boundaries between the two, it is often hard to separate where humanitarianism ends and development starts. Official Development Assistance: https://www.oecd.org/dac/financing-sustainable-development/development-finance-standards/official-development-assistance.htm and https://www.oecd.org/dac/financing-sustainable

-development/development-finance-data/ODA%202013%20Tables%20and %20Charts%20En.pdf

15 https://press.un.org/en/2023/sc15410.doc.htm
16 https://www.irisguard.com/who-we-are/about-us/
17 *Guardian*: 'Europe's Refugee Crisis is a major opportunity for businesses': https://www.theguardian.com/business/2015/sep/11/europes-refugee-crisis-is-a-major-opportunity-for-businesses; the *Financial Times*: 'Refugee camps are an untapped opportunity for the business sector': https://www.ft.com/content/e2d6588a-5042-11e8-b3ee-41e0209208ec. Following criticism in social media *The Financial Times* retracted the above headline; see here for details: https://twitter.com/FT/status/992801702939189249. See also the following post by UNHCR: 'For an increasing number of companies and investors, the refugee crisis is becoming both an economic and philanthropic opportunity': https://www.unhcr.org/news/stories/more-businesses-commit-helping-refugees-thrive-new-jobs-trainings-investment
18 Bill and Melissa Gates Foundation Facts: https://www.gatesfoundation.org/
19 Audience studies in the 1980s and 1990s was a rare moment of convergence, when the so-called critical theory and administrative traditions of research converged. For a discussion see Livingstone (1994).
20 https://www.irisguard.com/who-we-are/about-us/

2 *Biometric Infrastructures*

1 https://www.unhcr.org/jo/wp-content/uploads/sites/60/2022/02/1-Zaatari-Fact-Sheet-January-2022-final.pdf
2 World Food Programme, Building Blocks: https://innovation.wfp.org/project/building-blocks
3 'Biometrics investment continues as global market forecast to surpass $84 by 2026': https://www.biometricupdate.com/202110/biometrics-investment-continues-as-global-market-forecast-to-surpass-84b-by-2026
4 Kushi Baby is a wearable digital necklace with biometrically enabled access for tracking infants' immunisation records in India: https://www.khushibaby.org/case-details.html
5 UNHCR Global Report 2021: https://reporting.unhcr.org/globalreport2021
6 ProGres v4 users are assigned different levels of access depending on their role and authorization, so access can be very restricted. https://www.unhcr.org/registration-guidance/chapter3/registration-tools/
7 Office of the Inspector General, Internal Audit of SCOPE 2021: https://docs.wfp.org/api/documents/WFP-0000144800/download/?_ga=2.93348249.526843584.1695640963-1286107937.1695640963
8 The WFP Audit report highlighted the difficulty of correcting erroneous

entries and the quality of data uploaded, which were often incomplete or inaccurate: https://www.thenewhumanitarian.org/news/2018/01/18/exclusive-audit-exposes-un-food-agency-s-poor-data-handling. A copy of the 2021 Audit report can be found here: https://docs.wfp.org/api/documents/WFP-0000144800/download/?_ga=2.93348249.526843584.1695640963-1286107937.1695640963

9 ICRC Biometrics Policy: https://www.icrc.org/en/document/icrc-biometrics-policy

10 ICRC Rules on Personal Data Protection: https://www.icrc.org/en/publication/4261-icrc-rules-on-personal-data-protection

11 Oxfam Biometric Policy: https://oxfam.app.box.com/v/OxfamBiometricPolicy

12 Eurodac 2022 Statistics: https://www.eulisa.europa.eu/Publications/Reports/Eurodac%20-%202022%20Statistics%20-%20report.pdf

13 https://www.icrc.org/en/document/cyber-attack-icrc-what-we-know

14 https://www.hrw.org/news/2022/03/30/new-evidence-biometric-data-systems-imperil-afghans

15 https://www.thenewhumanitarian.org/investigation/2020/01/29/united-nations-cyber-attack

16 Convention on the Privileges and Immunities of the United Nations. New York, 13 February 1946: https://treaties.un.org/doc/Treaties/1946/12/19461214%2010-17%20PM/Ch_III_1p.pdf

17 https://www.hrw.org/news/2021/06/15/un-shared-rohingya-data-without-informed-consent

18 https://gdpr.eu/gdpr-consent-requirements/

19 https://redfish.media/cashing-in-on-crisis-the-refugee-eye-scan-experiment/

20 https://www.unhcr.org/jo/wp-content/uploads/sites/60/2022/02/1-Zaatari-Fact-Sheet-January-2022-final.pdf

21 https://www.thenewhumanitarian.org/news-feature/2022/11/22/Syrians-Jordan-Zaatari-camp-food-crisis-hunger-child-labour

22 Iazzolino in his ethnography of the Kakuma refugee camp in Kenya has also observed that biometrics restrict mobility (2021).

23 https://redfish.media/cashing-in-on-crisis-the-refugee-eye-scan-experiment/

3 Extracting Data and the Illusion of Accountability

1 Feedback surveys have become the standard method of establishing accountability in several sectors, including the UK Higher Education System, where the National Student Survey is used as the key measurement of student satisfaction, with subsequent implications for university ranking and funding.

2 Interagency feedback mechanisms were established to counter this but, as we shall see later in the chapter, not always successfully.

3 Dolores' story is also mentioned in an earlier publication (Madianou, Ong, Longboan and Cornelio, 2016).
4 This builds on earlier slogans about the Philippines being the 'texting capital of the world' (Pertierra et al., 2002).
5 In 2013, mobile teledensity in the Philippines was 103 per cent. In 2023 it was 144 per cent: https://data.worldbank.org/indicator/IT.CEL.SETS.P2?locations=PH. These figures do not mean that everyone in the Philippines has a mobile phone connection. Several users will have multiple connections (separate for work and personal use, for instance), while many people remain without mobile phones. Mobile teledensity does not account for regional or class differentiation. In the regions most affected by Typhoon Haiyan, which are some of the most deprived in the Philippines, mobile teledensity is lower than in the capital region. Like all metrics, mobile teledensity conceals inequalities within countries.
6 In their study of humanitarian feedback processes in Myanmar, Hilhorst et al. (2021) also observed the posting of 'thank you' messages.
7 World Vision's database listed over 3,000 feedback entries from suggestion boxes and SMS compared to only around 700 from help desks and community consultations over a period of one year (Buchanan-Smith, Ong and Routley, 2015).
8 There have been several sex abuse scandals involving staff of aid agencies. In 2018 two scandals involving Oxfam UK and Save the Children UK came to light, also in the context of the global #metoo campaigns. But the problem has deeper roots in what is a very male-dominated sector: https://odihpn.org/publication/tackling-sexual-exploitation-and-abuse-by-aid-workers-what-has-changed-20-years-on/
9 https://myanmar.unfpa.org/en/news/census-report-half-million-young-people-cannot-read-or-write
10 Inter-Sector Coordination Group (2017) *Situation Report: Rohingya Refugee Crisis*. Cox's Bazar: ISCG. 31 December 2017.
11 In the Philippines, like several countries in the global South, mobile telephone plans are usually pre-paid.
12 The World Bank lists the mobile cellular subscriptions (per 100 people) among the least developed countries: https://data.worldbank.org/indicator/IT.CEL.SETS.P2?end=2022&locations=XL&start=2022&view=map
13 https://www.who.int/emergencies/situations/ebola-%C3%A9quateur-province-democratic-republic-of-the-congo-2022
14 https://www.unhcr.org/uk/emergencies/dr-congo-emergency
15 https://www.unocha.org/haiti
16 https://www.unrefugees.org/emergencies/south-sudan/

17 Introduction to *Tawasul* chatbot: https://www.etcluster.org/sites/default/files/documents/Tawasul%20Chatbot_Libya.pdf
18 Humanitarian chatbot: https://www.wfp.org/stories/humanitarian-chatbot-how-tech-bridges-gap-between-people-and-assistance-they-need-ukraine

4 Surreptitious Experimentation: Enchantment, Coloniality and Control

1 *The New Yorker*, 'The chatbot will see you now': https://www.newyorker.com/tech/annals-of-technology/the-chatbot-will-see-you-now
2 The Karim chatbot example is discussed in an earlier publication (Madianou, 2021).
3 https://openai.com/
4 The Techfugees Foundation is a Private Limited Company registered as a software development business: https://find-and-update.company-information.service.gov.uk/company/10085961
5 'Welcome to the bootcamp': https://wfpinnovation.medium.com/welcome-to-bootcamp-687ff6a3de2
6 'Welcome to the bootcamp': https://wfpinnovation.medium.com/welcome-to-bootcamp-687ff6a3de2
7 Speaking at the Techfugees Global Summit in 2017: https://www.youtube.com/watch?v=O1NAIWLCzyk
8 WFP Hydroponics: https://innovation.wfp.org/project/h2grow-hydroponics
9 *The New Yorker*, 'The chatbot will see you now': https://www.newyorker.com/tech/annals-of-technology/the-chatbot-will-see-you-now; *Guardian*, 'Karim the AI delivers psychological support to Syrian refugees': https://www.theguardian.com/technology/2016/mar/22/karim-the-ai-delivers-psychological-support-to-syrian-refugees
10 https://www.irisguard.com/who-we-are/about-us/
11 https://data.humdata.org/
12 An example of a humanitarian chatbot that is only available in English is the WFP's CHITCHAT bot, which was piloted in the Kakuma refugee camp in Kenya. Defending that decision, the design team argued that most of their target users already spoke English. However, communication in English excludes older users, who were also experiencing additional access issues. Translating the bot into the languages (Dinka, Kiswahili or Somali) spoken in the western Kenya refugee camps is a vital step towards inclusivity.
13 It is worth noting that in 2023 the US National Eating Disorders Association suspended 'Tess' for giving harmful advice to patients: https://www.wired.com/story/tessa-chatbot-suspended/#:~:text=A%20nonprofit%20has%20suspended%20the,that%20could%20exacerbate%20eating%20disorders

14 'Zuckerberg to the UN: The Internet belongs to everyone': https://www.wired.com/2015/09/zuckerberg-to-un-internet-belongs-to-everyone/

15 Building Blocks was a project of the World Food Programme's Innovation Accelerator, which was launched in 2015 to pilot new solutions and scale innovations that address world hunger. The WFP Innovation Accelerator is one of the UN Agencies' innovation labs. UNHCR's Innovation Lab was launched in 2012, while UNICEF formalized its innovation efforts in 2007 (Maitland, 2018). Humanitarian NGOs also have innovation labs. IRC's Airbel Impact lab is one of the most active ones. Technological pilots take place outside the sphere of these formal organizations. Small NGOs and private companies also run their own pilots, which are not always named or recognized as such. Entrepreneurs and digital developers are also involved through the trial of apps or other untested innovations.

16 https://aiforgood.itu.int/

5 The Humanitarian Machine: Automating Harm

1 UNHCR, 'Afghan Recyclers under scrutiny of new technology': https://www.unhcr.org/news/afghan-recyclers-under-scrutiny-new-technology

2 https://www.wfp.org/stories/wfp-glance

3 The Alibaba Foundation is the non-profit wing of the Alibaba group, a Chinese electronic retail and technology company.

4 https://hungermap.wfp.org/

5 https://hungermap.wfp.org/

6 During March 2020 I interacted with 'Hunger Map Live' on the following dates: 9, 11, 16 and 23 March. When using the platform, I searched the profile of several countries, including Iran, which was one of the first countries to experience the first wave of Covid-19 in February and March 2020. My searches revealed that Covid-19 data in relation to Iran were available on 23 March 2020, several weeks after the first cases were recorded by the country: https://www.statista.com/chart/21100/coronavirus-in-iran/

7 https://www.unrefugees.org/news/inside-the-worlds-five-largest-refugee-camps/

8 UNHCR and WFP, 'Review of Targeting of Cash and Food Assistance for Syrian Refugees in Lebanon, Jordan and Egypt' (2015): https://www.calpnetwork.org/wp-content/uploads/2020/01/erc-targeting-cash-and-food-assistance-web.pdf

9 'Cash Transfer Programming for Syrian Refugees: Lessons Learned on Vulnerability, Targeting, and Protection from the Danish Refugee Council's E-Voucher Intervention in Southern Turkey': https://reliefweb.int/report/turkey/cash-transfer-programming-syrian-refugees-lessons-learned-vulnerability-targeting-and

10 Using local government units for aid distribution was one of the methods used in Tacloban. Another common method was geographic distribution (i.e., targeting assistance to one geographic area over others on the basis of criteria, such as material damage).
11 UNHCR Project Jetson: https://jetson.unhcr.org/#
12 One example is the IRC Placement Algorithm: https://airbel.rescue.org/projects/placement-algorithm/
13 The IRC Placement Algorithm was abandoned even before the prototyping phase: https://airbel.rescue.org/projects/placement-algorithm/
14 Privacy International has conducted an excellent investigation on the double registration in Kenya: https://privacyinternational.org/video/4412/when-id-leaves-you-without-identity-case-double-registration-kenya
15 Recall that here are different policies on biometric data retention. ICRC's policy mandates deletion, while UNHCR keeps all biometric data. See Chapter 3 for discussion.
16 'When ID leaves you without identity': https://privacyinternational.org/video/4412/when-id-leaves-you-without-identity-case-double-registration-kenya
17 https://interagencystandingcommittee.org/grand-bargain

6 Mundane Resistance: Contesting Technocolonialism in Everyday Life

1 While the Arab Spring was popularly framed as a 'revolution' the degree to which social change was achieved remains questionable.
2 I am aware of the recent controversy surrounding James Scott, following his own revelations that he had been contributing to CIA (the US Central Intelligence Agency) reports on the student movement in Myanmar before he began his fieldwork. I have decided to include his work here as the intellectual contribution remains very valuable, despite the revelation about this early period of his life. Readers may want to make up their own minds by reading the whole interview by Scott, where he discusses his role: https://digicoll.lib.berkeley.edu/record/219393?ln=en
3 'Moria migrants: fire destroys Greek camp leaving 13,000 without shelter': https://www.bbc.co.uk/news/world-europe-54082201
4 UNHCR Za'atari camp factsheet (2022): https://www.unhcr.org/jo/wp-content/uploads/sites/60/2022/02/1-Zaatari-Fact-Sheet-January-2022-final.pdf
5 The quotes from Carol and Dina in this section have appeared in an earlier publication (Madianou, Longboan and Ong, 2015).
6 For a discussion on the mediation and remediation of witnessing see Chouliaraki, 2013.
7 'Fire in Moria Refugee Camp': https://forensic-architecture.org/investigation/fire-in-moria-refugee-camp

8 'Fire in Moria Refugee Camp': https://forensic-architecture.org/investigation/fire-in-moria-refugee-camp
9 At the time of writing the appeal trial is scheduled for 2024.
10 'The Left-to-Die Boat': https://forensic-architecture.org/investigation/the-left-to-die-boat
11 https://xpmethod.columbia.edu/torn-apart/volume/1/
12 https://www.hrw.org/news/2023/07/11/you-dont-need-demand-sensitive-biometric-data-give-aid-ukraine-response-shows-how#:~:text=Digital%20literacy%20is%20high%20in,as%20part%20of%20their%20response
13 There was some basis to these fears. Some participants claimed that they had received warnings from local government officials about exclusion from aid if they attended protest events (Madianou, Longboan and Ong, 2015).

Conclusion

1 https://www.irisguard.com/who-we-are/about-us/
2 https://www.unocha.org/publications/report/occupied-palestinian-territory/humanitarian-access-snapshot-gaza-strip-1-31-march-2024
3 Both drones and satellite imagery are used in humanitarian operations; for example, to map damage during disasters. As my interlocutors have mentioned in our conversations over the past ten years, mapping using satellite imagery is easier for countries in the northern hemisphere, as there are more satellites than over the southern hemisphere. In other words, the global North has data, but the global South has statistical models – predictions. Similarly, as the technology of drones (UAVs) has improved and they fly longer distances, they give rise to conspiracy theories among local communities, who have no understanding of the purpose of these programmes.
4 For a discussion on a care-based approach to disaster see Ellcessor (2022); see also work by the care collective (Chatzidakis et al., 2020).
5 May First Movement Technology: https://mayfirst.coop/en/
6 https://www.computerweekly.com/news/366580074/Greek-government-fined-over-AI-surveillance-in-refugee-camps see also Homo Digitalis: https://homodigitalis.gr
7 https://www.thenewhumanitarian.org/opinion/2023/07/11/you-dont-need-demand-sensitive-biometric-data-give-aid-ukraine-response-shows

References

Achiume, T. (2021) 'Digital racial borders', *American Journal of International Law*, 115: 333–8.
Adams, V. (2013) *Markets of Sorrow, Labors of Faith: New Orleans in the Wake of Katrina*. Durham: Duke University Press.
Agamben, G. (1998) *Homo Sacer: Sovereign Power and Bare Life*, trans. Daniel Heller-Roazen. Stanford: Stanford University Press.
Agier, M. (2011) *Managing the Undesirables: Refugee Camps and Humanitarian Government*. Cambridge: Polity.
Ajana, B. (2013) *Governing Through Biometrics: The Biopolitics of Identity*. Basingstoke: Palgrave Macmillan.
ALNAP (2022) 'The state of the humanitarian system', ALNAP Study. London: ALNAP/ODI.
Ames, M. G. (2019) *The Charisma Machine: The Life, Death, and Legacy of One Laptop per Child*. Cambridge, MA: MIT Press.
Amoore, L. (2006) 'Biometric borders: Governing mobilities in the war on terror', *Political Geography*, 25(3): 336–51. https://doi.org/10.1016/j.polgeo.2006.02.001
Anderson, C. (2004) *Legible Bodies: Race, Criminality and Colonialism in South Asia*. Oxford: Berg.
Anderson, W. (2006) *Colonial Pathologies. American Tropical Medicine, Race and Hygiene in the Philippines*. Durham: Duke University Press.
Andersson, R. (2014) *Illegality Inc. Clandestine Migration and the Business of Bordering Europe*. Berkeley: University of California Press.
Andrejevic, M. and Gates, K. (2014) 'Big data surveillance: An introduction', *Surveillance and Society*, 12(2): 185–96.
Anwar, M. A. and Graham, M. (2020a) 'Digital labour at economic margins: African workers and the global information economy', *Review of African Political Economy*, 47(163): 95–105. doi: 10.1080/03056244.2020.1728243
Anwar, M. A. and Graham, M. (2020b) 'Hidden transcripts of the gig economy: Labour agency and the new art of resistance among African gig workers', *Environment and Planning A: Economy and Space*, 52(7): 1269–91.
Aouragh, M. and Chakravartty, P. (2016) 'Infrastructures of Empire: Towards

a critical geopolitics of media and information studies', *Media, Culture & Society*, 38(4): 559–75.

Appadurai, A. (1993) 'Number in the postcolonial imagination', in Breckenridge, C. and van der Veer, P. (eds), *Orientalism and the Postcolonial Predicament*. Philadelphia: University of Pennsylvania Press, pp. 314–39.

Aradau, C. (2022) 'Experimentality, surplus data and the politics of debilitation in borderzones', *Geopolitics*, 27(1): 26–46. https://doi.org/10.1080/14650045.2020.1853103

Aradau, C. and Blanke, T. (2017) 'Governing others: Anomaly and the algorithmic subject of security', *European Journal of International Security*, 3(1): 1–21. doi:10.1017/eis.2017.14

Arora, P. (2019) *The Next Billion Users: Digital Life Beyond the West*. Cambridge, MA: Harvard University Press.

Asad, T. (1994) 'Ethnographic representation, statistics and modern power', *Social Research*, 61(1): 55–88.

Bakardjieva, M. (2015) 'Rationalizing sociality: An unfinished script for socialbots', *The Information Society*, 31(3): 244–56. doi: 10.1080/01972243.2015.1020197

Banet-Weiser, S. and Higgins, K.C. (2023) *Believability: Sexual Violence, Media, and the Politics of Doubt*. Cambridge: Polity.

Barassi, V. (2021) *Child, Data, Citizen. How Tech Companies Are Profiling Us Before Birth*. Cambridge, MA: MIT Press.

Barbrook, R. and Cameron, A. (1996) *The Californian Ideology: Science as Culture*, 6(1): 44–72. doi: 10.1080/09505439609526455

Barnett, M. (2011) *Empire of Humanity. A History of Humanitarianism*. Ithaca and London: Cornell University Press.

Barnett, M. (2022) 'Humanitarianism's new business model', *Public Anthropologist*, 4(2): 233–59. doi: https://doi.org/10.1163/25891715-bja10039

Barnett, M. and Weiss, T. G. (eds) (2008) *Humanitarianism in Question*. Ithaca, NY: Cornell University Press.

Bhattacharyya, G. (2018) *Rethinking Racial Capitalism: Questions of Reproduction and Survival*. London: Rowman and Littlefield International.

Barocas, S. and Nissenbaum, H. (2014) 'Big Data's end run around anonymity and consent', in Lane, J., Stodden, V., Bender, S. and Nissenbaum, H. (eds), *Privacy, Big Data and the Public Good: Frameworks for Engagement*. Cambridge: Cambridge University Press, pp. 44–75.

Bauman, Z. (1989) *Modernity and the Holocaust*. Cambridge: Polity.

Bayat, A. (2013) *Life as Politics: How Ordinary People Change the Middle East*. Stanford: Stanford University Press.

Baym, N. (2015) *Personal Connections in a Digital Age*. 2nd edn. Cambridge: Polity.

Benjamin, R. (2019) *Race After Technology*. Cambridge: Polity.

Benjamin, R. (2022) *Viral Justice: How to Grow the World We Want*. Princeton: Princeton University Press.

Bhabha, H. K. (1994) *The Location of Culture*. London: Routledge.

Bhambra G. (2017) 'The current crisis of Europe: Refugees, colonialism, and the limits of cosmopolitanism', *European Law Journal*, 25: 395–40.

Bigo, D. (2002) 'Security and Immigration: Toward a critique of the government of unease', *Alternatives*, 27: 63–92.

Bishop, M. and Green, M. (2008) *Philanthrocapitalism: How the Rich Can Save the World and Why We Should Let Them*. London: A&C Black.

Bloch, M. and Parry, J. (1989) *Money and the Morality of Exchange*. Cambridge: Cambridge University Press.

Bolter, J. D. and Grusin, R. (2000) *Remediation*. Cambridge, MA: MIT Press.

Bowker, G. and Star, S. L. (1999) *Sorting Things Out: Classification and its Consequences*. Cambridge, MA: MIT Press.

Bowyer, K., and Burge, M. (eds) (2016) *Handbook of Iris Recognition*. London: Springer.

Bowyer, K., Hollingsworth, K. and Flynn, P. (2008) 'Image understanding for iris biometrics: A survey', *Computer Vision and Image Understanding*, 110 (2): 281–307.

Brauman, R. (2000) *L'Action Humanitaire*. Paris: Flammarion.

Breckenridge, K. (2014) *Biometric State: The Global Politics of Identification and Surveillance in South Africa, 1850 to the Present*. Cambridge: Cambridge University Press.

Broussard, M. (2018) *Artificial Unintelligence: How Computers Misunderstand the World*. Cambridge, MA: MIT Press.

Browne, S. (2015) *Dark Matters: On the Surveillance of Blackness*. Durham: Duke University Press.

Buchanan-Smith, M., Ong, J. and Routley, S. (2015) 'Who's listening? Accountability to affected people in the Haiyan response.' London: Plan International.

Bucher, T. and Helmond, A. (2017) 'The affordances of social media platforms', in Burgess, J. Poell, T. and Marwick, A. (eds), *The Sage Handbook of Social Media*. London: Sage, pp. 233–53.

Buolamwini J. and Gebru, T. (2018) 'Gender shades: Intersectional accuracy disparities in commercial gender classification', *Proceedings of Machine Learning*, 81: 1–15.

Burns, R. (2019) 'New frontiers of philanthro-capitalism', *Antipode*, 51(4): 1101–22.

Calhoun, C. (2008) 'The imperative to reduce suffering', in Barnett, M. and Weiss, T. G. (eds), *Humanitarianism in Question*. Ithaca, NY: Cornell University Press.

Caliskan, A., Bryson, J. and Narayanan, A. (2017) 'Semantics derived automatically from language corpora contain human-like biases.' *Science*, 356 (6334): 183–6.

Cammaerts, B., Bruter, M., Banaji, S., Harrison, S. and Anstead, N. (2014) 'The myth of youth apathy: Young Europeans' critical attitudes toward democratic life', *American Behavioral Scientist*, 58(5): 645–66.

Cannell, F. (1999) *Power and Intimacy in the Christian Philippines*. Cambridge: Cambridge University Press.

Carmi, E. (2021) 'A feminist critique to digital consent', *Seminar.net*, 17(2). doi: 10.5577/seminar.4291

Castells, M. (2012) *Networks of Outrage and Hope: Social Movements in the Internet Age*. Cambridge: Polity.

Charlson, F., von Ommeren, M., Flaxman, A., Cornett, J., Whiteford, H. and Saxena, S. (2019) 'WHO prevalence estimates of mental disorders in conflict settings: A systematic review and meta-analysis', *The Lancet*, 394(10194): 240–8.

Chatzidakis, A., Hakim, J., Littler, J., Rottenberg, C. and Segal, L. (2020) *The Care Manifesto: The Politics of Interdependence*. London: Verso.

Cheesman, M. (2022) 'Infrastructure, justice and humanitarianism', unpublished DPhil thesis, University of Oxford.

Chimni, B. S. (2000) 'Globalisation, humanitarianism and the erosion of refugee protection', Refugee Studies Centre Working Paper. Oxford: Oxford University.

Chouliaraki, L. (2013) *The Ironic Spectator: Solidarity in the Age of Post-Humanitarianism*. Cambridge: Polity.

Chouliaraki, L. (2015) 'Digital witnessing in conflict zones: The politics of remediation', *Information, Communication & Society*, 18(11): 1362–77. doi: 10.1080/1369118X.2015.1070890

Chouliaraki, L. and Georgiou, M. (2022) *The Digital Border*. New York: New York University Press.

Chun, W. H. K. (2011) *Programmed Visions: Software and Memory*. Cambridge, MA: MIT Press.

Cidell, J. (2008). 'Challenging the contours: Critical cartography, local knowledge, and the public', *Environment and Planning A: Economy and Space*, 40(5): 1202–18. https://doi.org/10.1068/a38447

Clark, R. (2016) '"Hope in a hashtag": The discursive activism of #WhyIStayed', *Feminist Media Studies*, 16(5): 788–804. doi: 10.1080/14680777.2016.1138235

Clarke, K. (2018) 'When do the dispossessed protest? Informal leadership and mobilization in Syrian refugee camps', *Perspectives on Politics*. Cambridge: Cambridge University Press, 16(3): 617–33. doi: 10.1017/S153759271800102O.

Clayton, S. (2020) *The New Internationalists*. London: Goldsmiths Press.

Cole, S. A. (2001) *Suspect Identities: A History of Fingerprinting and Criminal Identification*. Cambridge, MA: Harvard University Press.

Cooke, B. and Khothari, U. (eds) (2001) *Participation: The New Tyranny?* London: Zed Books.

Coppi, G. and Fast, L. (2019) 'Blockchain and distributed ledger technologies in the humanitarian sector', HPG Report. London: Humanitarian Partnership Group / ODI.

Costanza-Chock, S. (2020) *Design Justice: Community-Led Practices to Build the Worlds We Need*. Cambridge, MA: MIT Press.

Couldry, N. and Mejias, U. (2019) *The Costs of Connection*. Stanford: Stanford University Press.

Cowen, D. (2020) 'Following the infrastructures of empire: Notes on cities, settler colonialism, and method', *Urban Geography*, 41(4): 469–86. doi: 10.1080/02723638.2019.1677990

Craig, C., Cave, S., Dihal, K., Dillon, S., Montgomery, J., Singler, B. and Taylor, L. (2018) *Portrayals and Perceptions of AI and Why They Matter*. London: The Royal Society.

Crawford, K. (2021) *Atlas of AI. Power, Politics and the Planetary Costs of Artificial Intelligence*. New Haven: Yale University Press.

Curato, N. (2019) *Democracy in a Time of Misery*. Oxford: Oxford University Press.

Curato, N., Ong, J. and Longboan, L. (2016) 'Protest as interruption of the disaster imaginary', in Rovisco, M. and Ong, J. (eds), *Taking the Square: Mediated Dissent and Occupations of Public Space*. London: Rowman and Littlefield, pp. 75–96.

Cusicanqui, S. R. (2012) 'Ch'ixinakax utxiwa: A reflection on the practices and discourses of decolonization', *South Atlantic Quarterly*, 111(1): 95–109. doi:10.1215/00382876-1472612

Das, V. (2007) *Life and Words: Violence and the Descent into the Ordinary*. Berkeley: University of California Press.

Death, C. (2010) 'Counterconducts: A Foucauldian analysis of protest', *Social Movement Studies*, 9(3): 235–51.

De Certeau, M. (1988) *The Practice of Everyday Life*. Berkeley: University of California Press.

De Genova N. (2016) 'The European question: Migration, race and postcoloniality in Europe', *Social Text*, 34(3): 75–102.

De Kom, A. (2022 [1934]) *We the Slaves of Suriname*. Cambridge: Polity.

De Waal, A. (1997) *Famine Crimes: Politics and the Disaster Relief Industry in Africa*. London: James Currey Publishers.

Dencik, L., Hintz, A., Redden, J. and Treré, E. (2022) *Data Justice*. London: Sage.

Diepeveen, S., Bryant, J., Mohamud, F. et al. (2022) 'Data Sharing and Third-Party Monitoring in Humanitarian Response', HPG Working Paper. London: ODI.

Digital Impact Alliance (DIAL) (2018) 'Foodbot and the AIDA Chatbot builder: Case study'. https://digitalimpactalliance.org/wp-content/uploads/2019/03/mVAM.pdf

Dijstelbloem, H. (2021) *Borders as Infrastructure: The Technopolitics of Border Control*. Cambridge, MA: MIT Press.

Dijstelbloem, H. and Broeders, D. (2015) 'Border surveillance, mobility management and the shaping of non-publics in Europe', *European Journal of Social Theory*, 18(1): 21–38.

Donini, A. (2008) 'Through a glass, darkly: Humanitarianism and Empire', in Gunewardena, N. and Schuller, M. (eds), *Capitalising on Catastrophe: Neoliberal Strategies in Disaster Reconstruction*. Lanham: Altamira Press.

Du Bois, W. E. B. (1903) *The Souls of Black Folk*. Reprint. Oxford: Oxford University Press, 2007.

Du Gay, P. (2000) *In Praise of Bureaucracy: Weber, Organization, Ethics*. London: Sage.

Duffield, M. R. and Hewitt, V. M. (eds) (2009) *Empire, Development and Colonialism: The Past in the Present*. Woodbridge: James Currey.

Dyer, R. (1997/2017) *White*. London: Routledge.

Edwards, P. N. (2002) 'Infrastructure and modernity: Force, time, and social organization in the history of sociotechnical systems', in Thomas J. Misa, Philip Brey and Andrew Feenberg (eds), *Modernity and Technology*. Cambridge, MA: MIT Press.

Eisenlohr, P. (2011) 'Introduction: What is a medium? Theologies, technologies and aspirations', *Social Anthropology*, 19(1): 1–5.

Ellcessor, E. (2022) *In Case of Emergency: How Technologies Mediate Crisis and Normalize Inequality*. New York: New York University Press.

Escobar, A. (1995) *Encountering Development*. Princeton: Princeton University Press.

Escobar, A. (2018) *Designs for the Pluriverse: Radical Independence, Autonomy and the Making of Worlds*. Durham: Duke University Press.

Eubanks, V. (2017) *Automating Inequality. How High-Tech Tools Profile, Police and Punish the Poor*. New York: St Martin's Press.
Fanon, F. (1952) *Black Skin, White Masks*. Reprint. London: Penguin, 2021.
Fanon, F. (1959) *A Dying Colonialism*. New York: Grove Press.
Fanon F. (1961) *The Wretched of the Earth*. Reprint. London: Penguin, 2001.
Farmer, P. (2004) 'An anthropology of structural violence', *Current Anthropology*, 45(3): 305–25. doi:10.1086/382250
Farmer, P. (2005) *Pathologies of Power*. Berkeley: University of California Press.
Farmer, P. et al. (2013) *Reimagining Global Health*. University of California Press.
Fassin, D. (2012) *Humanitarian Reason: A Moral History of the Present*. Berkeley: University of California Press.
Fassin, D. and Rechtman, R. (2009) *The Empire of Trauma: An Inquiry into the Condition of Victimhood*. Princeton: Princeton University Press.
Fejerskov, A. (2022) *The Global Lab: Inequality, Technology and the Experimental Movement*. Oxford: Oxford University Press.
Feldman, I. (2008) 'Refusing invisibility: Documentation and memorialization in Palestinian refugee claims', *Journal of Refugee Studies*, 21(4): 498–516. https://doi.org/10.1093/jrs/fen044
Fenton, N. (2016) *Digital, Political, Radical*. Cambridge: Polity.
Ferguson, J. (1994) *The Anti-Politics Machine: 'Development', Depoliticization, and Bureaucratic Power in Lesotho*. Cambridge: Cambridge University Press.
Finn, M. (2018) *Documenting Aftermath. Information Infrastructures in the Wake of Disasters*. Cambridge, MA: MIT Press.
Foucault, M. (1998) *The History of Sexuality, vol. 1: The Will to Knowledge*. London: Penguin.
Foucault, M. (2000) *Power: Essential Works of Michel Foucault 1954–1984*. New York: The New Press.
Freire, P. (1970) *Pedagogy of the Oppressed*. London: Penguin.
Galtung, J. (1969) 'Violence, peace and peace research', *Journal of Peace Research*, 6(3): 167–91. https://doi.org/10.1177/002234336900600301
Geber, T. (2016) 'Hackathons and refugees: We can do better', *The Engine Room Blog*: https://www.theengineroom.org/hackathons-and-refugees-we-can-do-better/
Gebru, T. (2020) 'Race and gender', in M. Dubber, F. Pasquale and S. Das (eds), *The Oxford Handbook of Ethics of AI*, pp. 253–70.
Gehl, R. W. and Bakardjieva, M. (2017) 'Socialbots and their friends', in Gehl, R. W. and Bakadjieva, M. (eds), *Socialbots and their Friends: Digital Media and the Automation of Sociality*. New York, NY: Routledge.
Gell, A. (1992) 'The technology of enchantment and the enchantment of

technology', in Coote, J. and A. Shelton (eds), *Anthropology, Art and Aesthetics*. Oxford: Clarendon, pp. 40–66.

Gillespie, M., Osseiran, S. and Cheesman, M. (2018) 'Syrian refugees and the digital passage to Europe: Smartphone infrastructures and affordances', *Social Media & Society*. https://doi.org/10.1177/2056305118764440

Gilligan, C. (1982) *In a Different Voice: Psychological Theory and Women's Development*. Cambridge, MA: Harvard University Press.

Gilmore, R. W. (2010) 'What is to be done?' *American Quarterly*, 63(2): 245–65.

Glenn, E. N. (2015) 'Settler colonialism as structure: A framework for comparative studies of US race and gender formation', *Sociology of Race and Ethnicity*, 1(1): 52–72. doi:10.1177/2332649214560440

Glissant, É. (1997) *Poetics of Relation*. University of Michigan Press.

Goffman, E. (1990) *The Presentation of Self in Everyday Life*. London: Penguin Books.

Goody, J. (1986) *The Logic of Writing and the Organization of Society*. Cambridge: Cambridge University Press.

Gopal, P. (2019) *Insurgent Empire: Anticolonial Resistance and British Dissent*. London: Verso.

Graham, S. (2010) *Disrupted Cities when Infrastructure Fails*. New York: London: Routledge.

Greene, J., Basilico, M. T., Kim, H. and Farmer, P. (2013) 'Colonial medicine and its legacies', in Farmer, P. et al. (eds), *Reimagining Global Health*. Berkeley: University of California Press, pp. 33–73.

Greiffenhagen, C., Xu, X., and Reeves, S. (2023) 'The work to make facial recognition work', *Proceedings of the ACM in Human Computer Interaction*, 7: 1–30. doi.org/10.1145/3579531

Gruenewald, T. and Witteborn, S. (2022) 'Feeling good: Humanitarian virtual reality film, emotional style and global citizenship', *Cultural Studies*, 36(1): 141–61. doi: 10.1080/09502386.2020.1761415

Gunkel, D. (2018). 'Ars ex machina: Rethinking responsibility in the age of creative machines', in Guzman, A. (ed.), *Human–Machine Communication*. New York: Peter Lang, pp. 221–36.

Gupta, A. (2012) *Red Tape: Bureaucracy, Structural Violence, and Poverty in India*. Durham: Duke University Press.

Guyatt, H., Della Rosa, F. and Spencer, J. (2016) 'Refugee vulnerability study, Kakuma, Kenya', Nairobi: Kinetrica, UNHCR and WFP.

Hacking, I. (1990) *The Taming of Chance*. Cambridge: Cambridge University Press.

Hacking, I. (2006) 'Making up people', *London Review of Books*, 28(16).

Haggerty, K. D. and Ericson, R. V. (2000) 'The surveillant assemblage', *British Journal of Sociology*, 51: 605–22. https://doi.org/10.1080/00071310020015280

Han, B. C. (2015) *The Transparency Society*. Stanford: Stanford University Press.

Hartmann, C., Rhoades, A. and Santo, J. (2014) 'Starting the conversation: Information, feedback and accountability through communications with communities in post-Typhoon Philippines', Geneva, Switzerland: International Organization for Migration.

Hegde, R. (2016) *Mediating Migration*. Cambridge: Polity.

Heller, C., Pezzani, L. and Situ Studio (2012) 'Forensic oceanography: Report on the "left-to-die" boat'. London: Centre for Research Architecture, Goldsmiths, University of London. https://content.forensic-architecture.org/wp-content/uploads/2019/06/FO-report.pdf

Henriksen, S. E. and Richey, L. A. (2022) 'Google's tech philanthropy: Capitalism and humanitarianism in the digital age', *Public Anthropologist*, 4(1): 21–50. https://doi.org/10.1163/25891715-bja10030

Herzfeld, M. (1992) *The Social Production of Indifference: Exploring the Symbolic Roots of Western Bureaucracy*. Oxford: Berg.

Herzfeld, M. (2002) 'The absence presence: Discourses of crypto-colonialism', *South Atlantic Quarterly*, 101(4): 899–926. https://www.muse.jhu.edu/article/39112

Higgs, E. (2010) 'Fingerprints and citizenship: The British state and the identification of pensioners in the interwar period', *History Workshop Journal*, 69: 52–67. http://www.jstor.org/stable/40646093

Higgs, E. (2011) *Identifying the English: A History of Personal Identification, 1500 to the Present*. London: Continuum.

Hilhorst, D., Melis, S., Mena, R. and van Voorst, R. (2021) 'Accountability in humanitarian action', *Refugee Survey Quarterly*, 40(4): 363–89. https://doi.org/10.1093/rsq/hdab015

Hill, C. (2022) 'Poetic resistance: Karen long-distance nationalism, rap music, and YouTube', *International Journal of Cultural Studies*, 25(1): 30–50. doi:10.1177/13678779211027179

Hochschild, A. (2019 [1998]) *King Leopold's Ghost. A Story of Greed, Terror and Heroism in Colonial Africa*. London: Picador.

Hollingsworth, K., Bowyer, K. W. and Flynn, P. J. (2008) 'Pupil dilation degrades iris biometric performance', *Computer Vision and Image Understanding*, 113: 150–7.

Horst, H. A. and Miller, D. (2006) *The Cell Phone: An Anthropology of Communication*. Oxford: Berg.

Human Rights Watch (2021) UN shared Rohingya Data without informed consent. https://www.hrw.org/news/2021/06/15/un-shared-rohingya-data-without-informed-consent

Iazzolino, G. (2021) 'Infrastructure of compassionate repression: Making sense of biometrics in Kakuma refugee camp', *Information Technology for Development*, 27(1): 111–28. doi: 10.1080/02681102.2020.1816881

International Committee of the Red Cross, The Engine Room and Block Party (2017) *Humanitarian Futures for Messaging Apps*. Geneva: ICRC.

Irani, L. (2015) 'Hackathons and the making of entrepreneurial citizenship', *Science, Technology, & Human Values*, 40: 799–824.

Irani, L. (2019) *Chasing Innovation: Making Entrepreneurial Citizens in Modern India*. Princeton: Princeton University Press.

Jackson, S. J., Bailey, M. and Welles, B. F. (2020) *#Hashtag Activism: Networks of Race and Gender Justice*. Cambridge, MA: MIT Press.

Jacobs, A. (2015) 'Pamati Kita: Let's listen together', *Humanitarian Exchange*, 23(16): 16–18.

Jacobsen, K. L. (2015) *The Politics of Humanitarian Technology: Good Intentions, Unintended Consequences and Insecurity*. London: Routledge.

Jacobsen, K. L. (2022) 'Biometric data flows and unintended consequences of counterterrorism', *International Review of the Red Cross*, 103: 619–52.

James, C. L. R. (1938) *The Black Jacobins: Toussaint L'Ouverture and the San Domingo Revolution*. Reprint. London: Penguin, 2022.

Juskalian R. (2018) 'Inside the Jordan refugee camp that runs on blockchain.' *MIT Technology Review*. https://www.technologyreview.com/s/610806/inside-the-jordan-refugee-camp-that-runs-on-blockchain/

Karasti, H. and Blomberg, J. (2018) 'Studying infrastructuring ethnographically', *Computer Supported Cooperative Work*, 27: 233–65. https://doi.org/10.1007/s10606-017-9296-7

Kaun, A. and Stiernstedt, F. (2023) *Prison Media: Incarceration and the Future of Work and Technology*. Cambridge, MA: MIT Press.

Kember, S. and Zylinska, J. (2012) *Life After New Media. Mediation as a Vital Process*. Cambridge, MA: MIT Press.

Kennedy, H. (2018) 'Living with data: Aligning data studies and data activism through a focus on everyday experiences of datafication', *Krisis: Journal for Contemporary Philosophy*, 2018(1): 18–30.

Kennedy, H., Hill, R. L., Aiello, G., and Allen, W. (2016) 'The work that visualisation conventions do', *Information, Communication & Society*, 19(6): 715–35. https://doi.org/10.1080/1369118X.2016.1153126

Khiabany, G. (2016) 'Refugee crisis, imperialism and the pitiless war on the poor', *Media Culture and Society*, 38(5): 755–62.

Klein, N. (2007) *The Shock Doctrine*. London: Penguin.

Knorr-Cetina, K. (1999) *Epistemic Cultures: How the Sciences Make Knowledge*. Cambridge, MA: Harvard University Press.

Kraidy, M.M. (2016) *The Naked Blogger of Cairo: Creative Insurgency in the Arab World*. Cambridge, MA: Harvard University Press.

Krause, M. (2014) *The Good Project*. Chicago: Chicago University Press.

Kukutai, T. and Taylor, J. (eds) (2016) *Indigenous Data Sovereignty: Towards an Agenda*. Canberra: Australian National University Press.

Kundnani, A. (2023) *What is Anti-Racism? Why it Means Anti-Capitalism*. London: Verso.

Kwet, M. (2019) 'Digital colonialism: US empire and the new imperialism in the global South', *Race & Class*, 60(4): 3–26.

Larkin, B. (2013) 'The politics and poetics of infrastructure', *Annual Review of Anthropology* 42(1): 327–43. doi:10.1146/annurev-anthro-092412-155522

Latonero, M. and Kift, P. (2018) 'On digital passages and borders: Refugees and the new infrastructure for movement and control', *Social Media & Society*, 4(1). https://doi.org/10.1177/2056305118764432

Latour, B. (1988) *The Pasteurization of France*. Cambridge, MA: Harvard University Press.

Latour, B. and Woolgar, S. (1986) *Laboratory Life: The Construction of Scientific Facts*. Princeton: Princeton University Press.

Lehuedé, S. (2024) 'The double helix of data extraction: Radicalising reflexivity in critical data studies', *Technology and Regulation*, 2024: 84–92. doi: 10.26116/techreg.2024.009

Lemberg-Pedersen, M. and Haioty, E. (2020) 'Re-assembling the surveillable refugee body in the era of data-craving', *Citizenship Studies*, 24(5): 607–24. https://doi.org/10.1080/13621025.2020.1784641

Lester, A. and Dussart, F. (2014) *Colonization and the Origins of Humanitarian Governance*. Cambridge: Cambridge University Press.

Leurs, K. and Smets, K. (2018) 'Five questions for digital migration studies: Learning from digital connectivity and forced migration in(to) Europe', *Social Media & Society*, 3: 1–16.

Li, T. (2007) *The Will to Improve: Governmentality, Development, and the Practice of Politics*. Durham: Duke University Press.

Lindtner, S. M. (2021) *Prototype Nation: China and the Contested Promise of Innovation*. Princeton: Princeton University Press.

Livingstone, S. (1994) 'The rise and fall of audience research: An old story with a new ending', in Levy, M. and Gurevitch, M. (eds), *Defining Media Studies: Reflections on the Future of the Field*. New York: Oxford University Press, pp. 247–54.

Lugones, M. (2007) 'Heterosexualism and the colonial/modern gender system', *Hypatia*, 22(1): 186–219. doi:10.1111/j.1527–2001.2007.tb01156.x

Lyon, D. (2015) *Surveillance After Snowden*. Cambridge: Polity.

Macari, A. H. (in preparation) 'The Empathy Machine', PhD thesis in progress. Goldsmiths, University of London.

Madianou, M. (2015) 'Digital inequality and second-order disasters: Social media in the Typhoon Haiyan recovery', *Social Media & Society*, 1(2). https://doi.org/10.1177/2056305115603386

Madianou, M. (2019a) 'Technocolonialism: Digital innovation and data practices in the humanitarian response to refugee crises', *Social Media & Society*, 5(3): 1–13. https://doi.org/10.1177/2056305119863146

Madianou M. (2019b) 'The biometric assemblage: Surveillance, experimentation, profit, and the measuring of refugee bodies', *Television & New Media*. 20(6): 581–99. doi:10.1177/1527476419857682

Madianou, M. (2020) 'A second-order disaster? Digital technologies during the Covid-19 pandemic', *Social Media and Society*, 6(3), https://doi.org/10.1177/2056305120948168

Madianou, M. (2021) 'Nonhuman humanitarianism: When 'AI for good' can be harmful', *Information, Communication & Society*, 24(6): 850–68. doi: 10.1080/1369118X.2021.1909100

Madianou, M. (2022) 'Polymedia life', *Pragmatics and Society*, 12(5): 857–64. doi: https://doi.org/10.1075/ps.00051.mad

Madianou, M. and Miller, D. (2012) *Migration and New Media Transnational Families and Polymedia*. London: Routledge.

Madianou, M., Longboan, L. and Ong, J. C. (2015) 'Finding a voice through humanitarian technologies? Communication technologies and participation in disaster recovery', *International Journal of Communication*, 9: 3020–38. https://ijoc.org/index.php/ijoc/article/view/4142/1468

Madianou, M., Ong, J. C., Longboan, L. and Cornelio, J. S. (2016) 'The appearance of accountability: Communication technologies and power asymmetries in humanitarian aid and disaster recovery', *Journal of Communication*, 66(6): 960–81. https://doi.org/10.1111/jcom.12258

Magnet, S. A. (2011) *When Biometrics Fail: Gender, Race and the Technology of Identity*. Durham: Duke University Press.

Maitland, C. F. (2018) 'Introduction', in Maitland, C. F. (ed.), *Digital Lifeline? ICTs for Refugees and Displaced Persons*. Cambridge, MA: MIT Press, pp. 1–14.

Malkki, L. H. (1995) *Purity and Exile: Violence, Memory, and National Cosmology among Hutu Refugees in Tanzania*. Chicago: University of Chicago Press.

Manyozo, L. (2012) *Media, Communication and Development*. New Delhi: Sage.

Marcus, G. E. (1995) 'Ethnography in/of the world system: The emergence of multi-sited ethnography', *Annual Review of Anthropology* 24(1): 95–117.

Marres, N. and Stark, D. (2020) 'Put to the test: For a new sociology of testing', *British Journal of Sociology*, 71: 423–43. https://doi.org/10.1111/1468-4446.12746

Marvin, C. (1990) *When Old Technologies Were New: Thinking About Electric Communication in the Late Nineteenth Century*. Oxford: Oxford University Press.

Marx, K. and Engels, F. (1970) 'Manifesto of the Communist Party', in Marx and Engels *Selected Works*. London: Lawrence and Wishart, pp. 31–95.

Masso, A. and Kasapoglu, T. (2020) 'Understanding power positions in a new digital landscape: Perceptions of Syrian refugees and data experts on relocation algorithm', *Information, Communication and Society*, 23(8): 1203–1219. doi: 10.1080/1369118X.2020.1739731

Maurer, B. (2006) 'The anthropology of money', *Annual Review of Anthropology* 35(1): 15–36.

Mazower, M. (2014) *No Enchanted Place: The End of Empire and the Ideological Origins of the United Nations*. Princeton: Princeton University Press.

Mazzarella, W. (2010) 'Beautiful balloon: The digital divide and the charisma of new media in India, *American Ethnologist*, 37(4): 783–804.

Mbembe, A. (2001) *On the Postcolony*. Berkeley: University of California Press.

Mbembe, A. (2017) *Critique of Black Reason*. Durham: Duke University Press.

McDonald, S. (2019) 'From space to supply chain: Humanitarian data governance', SSRN. https://ssrn.com/abstract=3436179 or http://dx.doi.org/10.2139/ssrn.3436179

McStay, A. (2013) 'I consent: An analysis of the Cookie Directive and its implications for UK behavioral advertising', *New Media & Society*, 15(4): 596–611. https://doi.org/10.1177/1461444812458434

Medrado, A. and Rega, I. (2023) *Media Activism, Artivism, and the Fight Against Marginalisation in the Global South*. London: Routledge.

Meier, P. (2015) *Digital Humanitarians: How Big Data is Changing the Face of Humanitarian Response*. Boca Raton, FL: CRC Press.

Melucci, A. (1989) *Nomads of the Present*. Philadelphia: Temple University Press.

Metcalfe, P. (2022) 'It's not a bug, it's a feature: Control and injustice in datafied borders', PhD thesis, Cardiff University. https://orca.cardiff.ac.uk/id/eprint/150192/

Metcalfe, P. and Dencik, L. (2019) 'The politics of big borders: Data (in)justice and the governance of refugees', *First Monday*, 24(4). doi: 10.5210/fm.v24i4.9934.

Meyer, B. (2013) 'Mediation and immediacy: Sensational forms, semiotic ideologies and the question of the medium', in Boddy, J. and Lambek, M. (eds), *A Companion to the Anthropology of Religion*. Malden, MA: Wiley, pp. 309–26.

Mezzadra, S. and Neilson, B. (2019) *The Politics of Operations: Excavating Contemporary Capitalism*. Durham: Duke University Press.

Mignolo, W. (2011) *The Darker Side of Western Modernity*. Durham: Duke University Press.

Mignolo, W. and Walsh, C. (2018) *On Decoloniality: Concepts, Analytics, Praxis*. Durham: Duke University Press.

Milan, S. (2020) 'Techno-solutionism and the standard human in the making of the COVID-19 pandemic', *Big Data & Society*, 7(2). https://doi.org/10.1177/2053951720966781

Milan, S. and Treré, E. (2019). 'Big data from the south(s): Beyond data universalism', *Television & New Media*, 20(4): 319–35. https://doi.org/10.1177/1527476419837739

Molnar, P. (2020) Technological testing grounds. Report. European digital rights and the refugee law lab. https://edri.org/wp-content/uploads/2020/11/Technological-Testing-Grounds.pdf

Monahan, T. (2010) *Surveillance in the Time of Insecurity*. New Brunswick: Rutgers University Press.

Mortensen, M. (2015) 'Connective witnessing: Reconfiguring the relationship between the individual and the collective', *Information, Communication & Society*, 18(11): 1393–1406. doi: 10.1080/1369118X.2015.1061574

Nanavati, S., Thieme, M. and Nanavati, R. (2002) *Biometrics: Identity Verification in a Networked World*. New York: John Wiley.

Napolitano, A. and EuroMed Rights (2023) *Artificial Intelligence: The New Frontier of the EU's Border Externalisation Strategy*. Copenhagen: EuroMed Rights.

Natale, S. (2019) 'If software is narrative: Joseph Weizenbaum, artificial intelligence and the biographies of ELIZA', *New Media & Society*, 21(3): 712–28.

Noble, S. (2018) *Algorithms of Oppression*. New York: New York University Press.

Nyamnjoh, F. B. (2017) 'Incompleteness: Frontier Africa and the currency of conviviality', *Journal of Asian and African Studies*, 52(3): 253–70. https://doi-org.gold.idm.oclc.org/10.1177/0021909615580867

Office of Internal Oversight Services (OIOS) (2016) Audit of the Biometric Identity Management System at the Office of the United Nations High Commissioner for Refugees. Report 2016/181. Geneva: United Nations.

Oyěwùmí, O. (1997) *The Invention of Women: Making an African Sense of Western Gender Discourses*. Minneapolis: University of Minnesota Press.

Ozkul, D. (2023) 'Automating immigration and asylum: The uses of new technologies in migration and asylum governance in Europe'. Oxford: Refugee Studies Centre, University of Oxford.

Papacharissi, Z. (2020) *Affective Publics: Sentiment, Technology, and Politics.* Oxford: Oxford: University Press.

Papadopoulos, D., Stephenson, N. and Tsianos, V. (2008) *Escape Routes: Control and Subversion in the Twenty-First Century.* London: Pluto.

Papadopoulos, D. and Tsianos, V. S. (2013) 'After citizenship: Autonomy of migration, organizational ontology and mobile commons', *Citizenship Studies*, 17(2): 178–96. doi: 10.1080/13621025.2013.780736

Parker, B. (2020) The cyber attack the UN tried to keep under wraps. The New Humanitarian. https://www.thenewhumanitarian.org/investigation/2020/01/29/united-nations-cyber-attack

Parks, L. (2012) 'Technostruggles and the satellite dish: A populist approach to infrastructure', in Bolin, G. (ed.) *Cultural Technologies: The Shaping of Culture in Media and Society.* London: Routledge, pp. 64–84.

Parreñas, R. S. (2001) *Servants of Globalisation.* Stanford: Stanford University Press.

Pasquale, F. (2015) *The Black Box Society.* Cambridge, MA: Harvard University Press.

Patterson, O. (2022) *The Sociology of Slavery: Black Society in Jamaica 1655–1838.* 2nd edn. Cambridge: Polity.

Pelizza, A. (2020) 'Processing alterity, enacting Europe: Migrant registration and identification as co-construction of individuals and polities', *Science, Technology, & Human Values*, 45(2): 262–88. https://doi.org/10.1177/0162243919827927

Pertierra, R., Ugarte, E., Pingol, A. and Dacanay, N. L. (2002) *Txting selves: Cellphones and Philippine Modernity.* Manila: De La Salle University Press.

Peters, J. D. (2015) *The Marvellous Clouds: Toward a Philosophy of Elemental Media.* Chicago: University of Chicago Press.

Petryna, A. (2009) *When Experiments Travel. Clinical Trials and the Global Search for Human Subjects.* Princeton: Princeton University Press.

Pink, S., Horst, H., Postill, J., Hjorth, L., Lewis, T. and Tacchi, J. (2016) *Digital Ethnography: Principles and Practice.* London: Sage.

Plantin, C., Lagoze, C., Edwards, P. N. and Sandvig, C. (2018) 'Infrastructure studies meet platform studies in the age of Google and Facebook', *New Media & Society*, 20(1): 293–310. https://doi.org/10.1177/1461444816661553

Plantin, C. and Punathambekar, A. (2019) 'Digital media infrastructures: Pipes, platforms and politics', *Media, Culture & Society*, 41(2): 163–74.

Ponzanesi, S. (2024) 'Posthumanitarianism and the crisis of empathy', in De

Medeiros, P. and Ponzanesi, S. (eds), *Postcolonial Theory and Crisis*. Berlin: De Gruyter, pp. 21–46.

Popper, K. R. (1959) *The Logic of Scientific Discovery*. London: Routledge.

Porter, T. M. (1995) *Trust in Numbers: The Pursuit of Objectivity in Science and Public Life*. Princeton: Princeton University Press.

Pugliese, J. (2010) *Biometrics: Bodies, Technologies, Biopolitics*. London: Routledge.

Quijano, A. (2000) 'Coloniality of power and eurocentrism in Latin America', *International Sociology*, 15(2): 215–32.

Rabinow, P. (1995) *French Modern: Norms and Forms of the Social Environment*. Chicago: University of Chicago Press.

Rajak, D. (2011) *In Good Company: An Anatomy of Corporate Social Responsibility*. Stanford: Stanford University Press.

Rao, U. and Nair, V. (2019) 'Aadhaar: Governing with biometrics, South Asia', *Journal of South Asian Studies*, 42(3): 469–81. doi: 10.1080/00856401.2019.1595343

Raymond, N., Scarnecchia, D. and Campo, S. (2017) 'Humanitarian data breaches: The real scandal is our collective inaction', IRIN, 8 December. https://www.irinnews.org/opinion/2017/12/08/humanitarian-data-breaches-real-scandal-our-collectiveinaction

Ricaurte, P. (2019) 'Data epistemologies, the coloniality of power, and resistance', *Television & New Media*, 20(4): 350–65.

Rieff, D. (2002) *A Bed for the Night: Humanitarianism in Crisis*. London: Vintage.

Risam, R. (2023) 'Connecting the data points: Refugee data narratives', in Vinh Nguyen and Evelyn Lê Espiritu Gandhi (eds), *Routledge Handbook of Refugee Studies*. London: Routledge, pp. 153–64.

Robinson, C. (2021) *Black Marxism: The Making of the Black Radical Tradition*. 3rd edn. London: Penguin.

Rosenberg, C. (2012) 'The international politics of vaccine testing in interwar Algiers', *American Historical Review*, 117(3): 671–97. doi:10.1086/ahr.117.3.671.

Ruppert, E. S. and Scheel, S. (2021) *Data Practices: Making Up a European People*. London: Goldsmiths Press.

Said, E. W. (1978) *Orientalism*. London: Penguin. Reprint. 2003.

Said, E. W. (1994) *Culture and Imperialism*. London: Vintage.

Sandvik, K. B. (2020) 'Wearables for something good: Aid, dataveillance and the production of children's digital bodies', *Information, Communication & Society*, 23(14): 2014–29.

Sandvik, K. B., Jacobsen, K. L., and McDonald, S. M. (2017) 'Do no harm: A taxonomy of the challenges of humanitarian experimentation', *International Review of the Red Cross*, 99(904): 319–44.

Santos, B. S. (2016) *Epistemologies of the South: Justice against Epistemicide*. London: Routledge.

Scheel, S. (2019) *Autonomy of Migration? Appropriating Mobility within Biometric Border Regimes*. London and New York: Routledge.

Schneider, N. (2022) 'Governable stacks against digital colonialism', *TripleC*, 20 (1): 19–36. https://doi.org/10.31269/triplec.v20i1.1281

Scott, J. (1985) *Weapons of the Weak: Everyday Forms of Peasant Resistance*. New Haven and London: Yale University Press.

Scott, J. (1990) *Domination and the Arts of Resistance: Hidden Transcripts*. New Haven and London: Yale University Press.

Scott, J. (1998) *Seeing like a State: How Certain Schemes to Improve the Human Condition Have Failed*. New Haven: Yale University Press.

Scott-Smith, T. (2016) 'Humanitarian neophilia: The "innovation turn" and its implications', *Third World Quarterly*, 37(12): 2229–51. doi: 10.1080/01436597.2016.1176856

Seaver, N. (2017) 'Algorithms as culture: Some tactics for the ethnography of algorithmic systems', *Big Data & Society*, 4(2). doi:10.1177/2053951717738104

Seuferling, P. and Leurs, K. (2021) 'Histories of humanitarian technophilia: How imaginaries of media technologies have shaped migration infrastructures', *Mobilities*, 16(5): 670–87. doi: 10.1080/17450101.2021.1960186

Shi, Z. R., Wang, C., and Fang, F. (2020) 'Artificial intelligence for social good: A survey.' ArXiv [preprint]. arXiv:2001.01818 [cs.CY].

Shoemaker, E., Currion, P. and Bon, B. (2018) *Identity at the Margins: Identification Systems for Refugees*. Farnham, Surrey: Caribou Digital Publishing.

Sims, C. (2017) *Disruptive Fixations: Social Reform and the Politics of Techno-Idealism*. Princeton: Princeton University Press.

Singh, R. and Jackson, S. (2021) 'Seeing as an infrastructure', *Proceedings of the ACM on Human–Computer Interaction*, Vol. 5, CSCW2, Article 315.

Singha, R. (2000) 'Settle, mobilize, verify: Identification practices in Colonial India', *Studies in History*, 16(2): 151–98.

Skinner, R. and Lester, A. (2012) 'Humanitarianism and Empire: New research agendas', *Journal of Imperial and Commonwealth History*, 40(5): 729–47. doi: 10.1080/03086534.2012.730828

Slim, H. (2015) *Humanitarian Ethics: A Guide to the Morality of Aid in War and Disaster*. Oxford: Oxford University Press.

Smith, L.T. (2021) *Decolonizing Methodologies: Research and Indigenous Peoples*. London: Bloomsbury Academic.

Sphere (2018) *The Sphere Handbook: Humanitarian Charter and Minimum Standards in Humanitarian Response*. Geneva: Sphere Association.

Spivak, G. C. (1999) *A Critique of Postcolonial Reason: Toward a History of the Vanishing Present*. Boston: Harvard University Press.
Spivak, G. C. (2010) 'Can the subaltern speak?' in Morris, R.C. (ed.), *Can the Subaltern Speak? Reflections on the History of an Idea*. New York: Columbia University Press, pp. 21–78.
Sreberny, A. and Khiabany, G. (2010) *Blogistan: The Internet and Politics in Iran*. London: I. B. Tauris.
Star, S. L. (1999) 'The ethnography of infrastructure', *American Behavioral Scientist*, 43(3): 377–91. doi: 10.1177/00027649921955326
Star, S. L. and Ruhleder, K. (1996) 'Steps toward an ecology of infrastructure: Design and access for large information spaces', *Information Systems Research*, 7(1): 111–34.
Starosielski, N. (2015) *The Undersea Network*. Durham: Duke University Press.
Stein, J. G. (2008) 'Humanitarian organizations: Accountable: Why, to whom, for what, and how?' in Barnett, M. and Weiss, T. (eds), *Humanitarianism in Question: Politics, Power, Ethics*. Ithaca: Cornell University Press, pp. 124–43.
Stoler A. L. (2008). 'Imperial debris: Reflections on ruins and ruination', *Cultural Anthropology*, 23: 191–219.
Stoler, A. L. (2016) *Duress: Imperial Durabilities in Our Times*. Durham: Duke University Press.
Strathern, M. (2000) *Audit Cultures*. London: Routledge.
Streeter, T. (2011) *The Net Effect: Romanticism, Capitalism, and the Internet*. New York: New York University Press.
Styers, R. (2004) *Making Magic: Religion, Magic, and Science in the Modern World*. Oxford: Oxford University Press.
Suchman, L. A. (2007) *Human–Machine Configurations. Plans and Situated Actions*. Cambridge: Cambridge University Press.
Suman, S. (2009) 'Putting knowledge in its place: Science, colonialism and the postcolonial', *Postcolonial Studies*, 12(4): 373–88.
Supp-Montgomerie, J. (2021) *When the Medium was the Mission: the Atlantic Telegraph and the Religious Origins of Network Culture*. New York: New York University Press.
Taylor, C. (2002) 'Modern social imaginaries', *Public Culture*, 14(1): 91–124.
Taylor, L. (2016) 'No place to hide? The ethics and analytics of tracking mobility using mobile phone data', *Environment and Planning D*, 34(2): 319–36.
Tazzioli, M. (2019) 'Refugees' debit cards, subjectivities, and data circuits: Financial-humanitarianism in the Greek migration laboratory', *International Political Sociology*, 13 (4): 92–408. https://doi.org/10.1093/ips/olz014

Terry, F. (2002) *Condemned to Repeat? The Paradox of Humanitarian Action*. Ithaca: Cornell University Press.

Thatcher, J., O'Sullivan, D. and Mahmoudi, D. (2017) 'Data colonialism through accumulation by dispossession: New metaphors for daily data', *Environment and Planning D: Society and Space*, 34(6): 990–1006.

The Engine Room, Tsui, Q., Perosa, T. and Singler, S. (2023). 'Biometrics in the humanitarian sector'. https://www.theengineroom.org/biometrics-humanitarian-sector-2023/

Thiong'o, N. wa (1986) *Decolonizing the Mind: The Politics of Language in African Literature*. London: J. Currey.

Thompson, J. B. (2021) *Book Wars: The Digital Revolution in Publishing*. Cambridge: Polity.

Ticktin, M. (2011) *Casualties of Care: Immigration and the Politics of Humanitarianism in France*. Berkeley: University of California Press.

Ticktin, M. (2016) 'Thinking beyond humanitarian borders', *Social Research*, 83 (2): 255–71.

Tilley, H. (2011) *Africa as a Living Laboratory: Empire, Development and the Problem of Scientific Knowledge, 1870–1950*. Chicago: University of Chicago Press.

Torres, A. T. (1987) 'The Filipina looks at herself: A review of women's studies in the Philippines', *Transactions of the National Academy of Science and Technology*, 9: 307–30.

Toupin, S. (2024) 'Shaping feminist artificial intelligence', *New Media & Society*, 26(1): 580–95. https://doi.org/10.1177/14614448221150776

Treré, E. and Bonini, T. (2022) 'Amplification, evasion, hijacking: Algorithms as repertoire for social movements and the struggle for visibility', *Social Movement Studies*. doi: 10.1080/14742837.2022.2143345

Tsing, A. (2009) 'Supply chains and the human condition', *Rethinking Marxism* 21(2): 148–76.

Tuck, E. and Yang, K. W. (2012) 'Decolonisation is not a metaphor', *Decolonization, Indigeneity, Education & Society*, 1(1): 1–40.

Tufekci, Z. (2017) *Twitter and Tear Gas: The Power and Fragility of Networked Protest*. New Haven: Yale University Press.

Tully, J. (2009) 'A Victorian ecological disaster: Imperialism, the Telegraph, and Gutta-Percha', *Journal of World History*, 20(4): 559–79. http://www.jstor.org/stable/40542850

Turing, A. M. (1950) 'Computing machinery and intelligence', *Mind*, LIX(236): 433–60. doi:10.1093/mind/lix.236.433

Turner, F. (2006) *From Counterculture to Cyberculture: Stewart Brand, the Whole*

Earth Network, and the Rise of Digital Utopianism. Chicago: University of Chicago Press.
United Nations High Commissioner for Refugees (UNHCR) (2002) Afghan "recyclers" under scrutiny of new technology. https://www.unhcr.org/news/latest/2002/10/3d9c57708/afghan-recyclers-under-scrutiny-new-technology.html
United Nations Office for the Coordination of Humanitarian Affairs (UNOCHA) (2013) *Humanitarianism in the Network Age*. OCHA policy and studies series. New York: OCHA.
Waisbord, S. (2008) 'The institutional challenges of participatory communication in international aid', *Social Identities*, 14(4): 505–22.
Walia, H. (2021) *Border and Rule: Global Migration, Capitalism, and the Rise of Racist Nationalism*. Chicago: Haymarket Books.
Weiss, T. G. (2013) *Humanitarian Business*. Cambridge: Polity.
Weitzberg, K. (2015) 'The unaccountable census: Colonial enumeration and its implications for the Somali people of Kenya', *Journal of African History*, 56(3): 409–28. http://www.jstor.org/stable/44509221
Weitzberg, K. (2020) 'Passing as a refugee', *Africa is a Country*, 10 November 2020. https://africasacountry.com/2020/11/passing-as-a-refugee
Weizenbaum, J. (1966) 'Eliza – a computer program for the study of natural language communication between man and machine', *Communications of the ACM*, 9(1): 36–45. doi:10.1145/365153.365168
Weizman, E. (ed.) (2014) *Forensis: The Architecture of Public Truth*. Berlin: Sternberg.
Wigley, B. (2015) 'Constructing a culture of accountability: Lessons from the Philippines', *Humanitarian Exchange*, 23: 13–16.
Winner, L. (1997) 'Technology today: Utopia or dystopia?', *Social Research*, 64(3): 989–1017.
Wolfe, P. (2006) 'Settler colonialism and the elimination of the Native', *Journal of Genocide Research*, 8(4): 387–409. doi:10.1080/14623520601056240
Younge, G. (2007) 'We used to think there was a Black community', *Guardian*, 8 November. https://www.theguardian.com/world/2007/nov/08/usa.gender
Zelizer, V. A. R. (1997) *The Social Meaning of Money: Pin Money, Paychecks, Poor Relief and Other Currencies*. Princeton: Princeton University Press.
Zuboff, S. (2019) *The Age of Surveillance Capitalism: The Fight for the Human Future at the New Frontier of Power*. London: Profile.

Index